Encyclopaedic Dictionary of Telecommunication

Vol. 2

Rajiv Shukla

RAJAT PUBLICATIONS

NEW DELHI - 110 002 (INDIA)

RAJAT PUBLICATIONS
4740/23, Ansari Road, Daryaganj
New Delhi- 110 002 (India)
Phones : +91-11-23267924, +91-9811484973
E-mail: rajat_publication@yahoo.com

Encyclopaedic Dictionary of Telecommunication

Edition 2025

ISBN 978-81-7880-307-4 (Set)

[The responsibility for facts stated opinion expressed or conclusions reached and plagiarism, if any, in this volume is entirely that of the Editor. The publisher bears no responsibility for them whatsoever.]

PRINTED IN INDIA

Published by Mrs. Seema Wasan for Rajat Publications, New Delhi and Printed at H.S. Offset Press, Delhi.

Preface

Telecommunications is the transmission of data and information between computers using a communications link such as a standard telephone line. Typically, a basic telecommunications system would consist of a computer or terminal on each end, communication equipment for sending and receiving data, and a communication channel connecting the two users. Appropriate communications software is also necessary to manage the transmission of data between computers. Some applications that rely on this communications technology include the following:

1. Electronic mail (e-mail) is a message transmitted from one person to another through computerized channels. Both the sender and receiver must have access to on-line services if they are not connected to the same network. E-mail is now one of the most frequently used types of telecommunication.

2. Facsimile (fax) equipment transmits a digitized exact image of a document over telephone lines. At the receiving end, the fax machine converts the digitized data back into its original form.

3. Voice mail is similar to an answering machine in that it permits a caller to leave a voice message in a voice mailbox. Messages are digitized so the caller's message can be stored on a disk.

4. Videoconferencing involves the use of computers, television cameras, and communications software and equipment. This equipment makes it possible to conduct electronic meetings while the participants are at different locations.

5. The Internet is a continuously evolving global network of computer networks that facilitates access to information on thousands of topics. The Internet is utilized by millions of people daily.

Actually, telecommunications is not a new concept. It began in the mid-1800s with the telegraph, whereby sounds were translated manually into words; then the telephone, developed in 1876, transmitted voices; and then the teletypewriter, developed in the early 1900s, was able to transmit the written word.

Since the 1960s, telecommunications development has been rapid and wide reaching. The development of dial modem technology accelerated the rate during the 1980s. Facsimile transmission also enjoyed rapid growth during this time. The 1990s have seen the greatest advancement in telecommunications. It is predicted that computing performance will double every eighteen months. In addition, it has been estimated that the power of the computer has doubled thirty-two times since World War II (With row, 1997). The rate of advancement in computer technology shows no signs of slowing.

Deregulation and new technology have created increased competition and widened the range of network services available throughout the world. This increase in telecommunication capabilities allows businesses to benefit from the information revolution in numerous ways, such as streamlining their inventories, increasing productivity, and identifying new markets.

True source for the latest terms and concepts used today in the field of telecommunications. The objective of this book is to provide a self-contained quick-reference to telecommunications jargon and facts in a clear concise manner.

Written and reviewed by industry expert, this dictionary cover topics spanning the telecommunications and information technology industry, including distribution design, insatllation, outside plant, and LAN and internetworking design. The dictionary is based on recognized codes, standards, and regulations, and include extensive glossaries.

—Editor

E & M signaling. In telephony, an arrangement that uses separate leads, called the "E" lead and "M" lead, for signaling and supervisory purposes. *Note 1:* The near end signals the far end by applying –48 vdc to the "M" lead, which results in a ground being applied to the far end's "E" lead. When –48 vdc is applied to the far end "M" lead, the near-end "E" lead is grounded. *Note 2:* The "E" originally stood for "ear," *i.e.*, when the near-end "E" lead was grounded, the far end was calling and "wanted your ear." The "M" originally stood for "mouth," because when the near-end wanted to call (*i.e.*, speak to) the far end, –48 vdc was applied to that lead.

earth. *See* ground.

Earth coverage (EC). In satellite communications, the coverage that occurs when the satellite-to-Earth beam is sufficiently wide to cover all of the surface of the Earth exposed to the satellite, *i.e.*, the footprint is as large as it can possibly be from a geographic standpoint.

earth electrode subsystem. A network of electrically interconnected rods, plates, mats, or grids, installed and connected, for the purpose of establishing a low-resistance contact with earth.

Earth exploration-satellite service. A radio-communication service between Earth stations and one or more space stations, which may include links between space stations, in which:

- information relating to the characteristics of the Earth and its natural phenomena is obtained from active sensors or passive sensors on Earth satellites;
- similar information is collected from airborne or Earth-based platforms;
- such information may be distributed to Earth stations within the system concerned;
- platform interrogation may be included.

This service may also include feeder links necessary for its operation.

earth ground. *See* ground.

Earth station. A station located either on the Earth's surface or within the major portion of the Earth's atmosphere and intended for communication:

- with one or more space stations; or
- with one or more stations of the same kind by means of one or more reflecting satellites or other objects in space.

Earth terminal. In a satellite link, one of the non-orbiting communications stations that receives, processes, and transmits signals between itself and a satellite. *Note:* Earth terminals may be at mobile, fixed, airborne, and waterborne Earth terminal complexes. *Synonym* satellite Earth terminal.

Earth terminal complex. In satellite communications systems, the assembly of equipment and facilities necessary to integrate an Earth terminal into a communications network. *Note:* The Earth terminal complex includes the Earth terminal and its support equipment and any required interconnect facilities and their support equipment. It does not include facilities at the site that are not necessary to establish and integrate the satellite links with the network.

EAS. *Abbreviation for* extended area service.

EBCDIC. *Acronym for* extended binary coded decimal interchange code. An 8-bit alphanumeric coded character set.

E-bend. A smooth change in the direction of the axis of a waveguide, throughout which the axis remains in a plane parallel to the direction of electric E-field (transverse) polarization. *Synonym* E-plane bend.

E_bN_0. *See* signal-to-noise ratio per bit.

EBS. *Abbreviation for* Emergency Broadcast System.

EC. *Abbreviation for* Earth coverage.

ECCM. *Abbreviation for* electronic counter-countermeasures.

echo. 1. A wave that has been reflected by a discontinuity in the propagation medium. 2. A wave that has been reflected or otherwise returned with sufficient magnitude and delay to be perceived. *Note 1:* Echoes are frequently measured in dB relative to the directly transmitted wave. *Note 2:* Echoes may be desirable (as in radar) or undesirable (as in telephone systems). 3. In computing, to print or display characters (a) as they are entered from an input device, (b) as instructions are executed, or (c) as retransmitted characters received from a remote terminal. 4. In computer graphics, the immediate notification of the current values provided by an input device to the operator at the display console.

echo area. *See* scattering cross section.

echo attenuation. In a communication circuit (4- or 2-wire) in which the two directions of transmission can be separated from each other, the attenuation of echo signals that return to the input of the circuit under consideration. *Note:* Echo attenuation is expressed as the ratio of the transmitted power to the received echo power in dB.

echo cancellation. In a system, the reduction of the power level of an echo or the elimination of an echo. *Note:* Echo cancellation is usually an active process in which echo signals are measured and canceled or eliminated by combining an inverted signal with the echo signal.

echo canceler. *See* echo suppressor.

echo check. A check to determine the integrity of transmission of data, whereby the received data are returned to the source for comparison with the originally transmitted data. *Synonym* loop check.

echo command. A programming language command that causes an echo response from a computer to be displayed on a monitor or printer for analysis or monitoring of the progress of processing.

echo effect. *See* ghost.

echo line. In computer systems, on a hard-output or display device, a line of information that verifies (reflects) data entered into the computer.

echoplex. An echo check used in public switched networks operating in the full-duplex transmission mode, *i.e.,* the two-way-simultaneous mode.

echo return loss. *See* return loss.

echo sounding. The measurement of the depth of a body of water or the distance to an object in a body of water by measuring the time it takes sound or electromagnetic waves of known velocity to reflect from the bottom of the water body or from the distant object. *Note:* In echo sounding, damped cw transmission is usually used.

echo suppressor. A device for connection to a two-way telephone circuit to attenuate echo signals in one direction caused by signals in the other direction.

ECM. *Abbreviation for* electronic countermeasures.

edge busyness. In a video display, distortion that is concentrated at or near the edge of objects, and that is categorized further by its temporal and spatial characteristics.

edge-emitting LED. An LED that has a physical structure superficially resembling that of an injection laser diode, operated below the lasing threshold and emitting incoherent light. *Note:* Edge-emitting LEDs have a relatively small beam divergence, and thus are capable of launching more optical power into a given fiber than are the conventional surface-emitting LEDs.

EDI. *Abbreviation for* electronic document interchange.

EDTV. *Abbreviation for* extended-definition television.

effective antenna gain contour (of a steerable satellite beam). An envelope of antenna gain contours resulting from moving the boresight of a steerable satellite beam along the limits of the effective boresight area.

effective boresight area (of a steerable satellite beam). An area on the surface of the Earth within which the boresight of a steerable satellite beam is pointed. There may be more than one unconnected effective boresight area to which a single steerable satellite beam can be pointed.

effective data transfer rate. The average number of units of data, such as bits, characters, blocks, or frames, transferred per unit time from a source and accepted as valid by a sink. *Note:* The effective data transfer rate is usually expressed in bits, characters, blocks, or frames per second. The effective data transfer rate may be averaged over a period of seconds, minutes, or hours.

effective Earth radius. The radius of a hypothetical Earth for which the distance to the radio horizon, assuming rectilinear propagation, is the same as that for the actual Earth with an assumed uniform vertical gradient of atmospheric refractive index. *Note:* For the standard atmosphere, the effective Earth radius is 4/3 that of the actual Earth radius.

effective height. 1. The height of the center of radiation of an antenna above the effective ground level. 2. In low-frequency applications involving loaded or nonloaded vertical antennas, the moment of the current distribution in the vertical section divided by the input current. *Note:* For an antenna with symmetrical current distribution, the center of radiation is the center of distribution. For an antenna with asymmetrical current distribution, the center of radiation is the center of current moments when viewed from points near the direction of maximum radiation.

effective input noise temperature. The source noise temperature in a two-port network or amplifier that will result in the same output noise power, when connected to a noise-free network or amplifier, as that of the actual network or amplifier connected to a noise-free source. *Note:* If F is the noise figure numeric and 290 K the standard noise temperature, then the effective noise temperature is given by $T_n = 290(F\mathrm{B}1)$.

effective isotropically radiated power (e.i.r.p.). The arithmetic product of (a) the power supplied to an antenna and (b) its gain.

effective mode volume. For an optical fiber, the square of the product of the diameter of the near-field pattern and the sine of the radiation angle of the far-field pattern. The diameter of the near-field radiation pattern is defined here as the full width at half maximum and the radiation angle at half maximum intensity. *Note:* Effective mode volume is proportional to the breadth of the relative distribution of power amongst the modes in a multimode fiber. It is not truly a spatial volume but rather an *"optical volume"* equal to the product of area and solid angle.

effective monopole radiated power (e.m.r.p.). The product of the power supplied to the antenna and its gain relative to a short vertical antenna in a given direction.

effective power. In alternating-current power transmission and distribution, the product of the rms voltage and amperage, *i.e.,* the apparent power, multiplied by the power factor, *i.e.,* the cosine of the phase angle between the voltage and the current. *Note:* Only effective power, *i.e.,* the actual power delivered to or consumed by the load, is expressed in watts. Apparent power is properly expressed only in volt-amperes, never watts. *Synonym* true power. *See figure on following page.*

effective radiated power (e.r.p.) (in a given direction). 1. The power supplied to an antenna multiplied by the antenna gain in a given direction. *Note 1:* If the direction is not specified, the direction of maximum gain is assumed. *Note 2:* The type of reference antenna must be specified. 2. The product of the power supplied to the antenna and its gain relative to a half-wave dipole in a given direction. *Note:* If the direction is not specified, the direction of maximum gain is assumed.

effective speed of transmission. *Synonym* effective transmission rate.

effective transmission rate. The rate at which information is processed by a transmission facility. *Note 1:* The effective transmission rate is calculated as (a) the measured number of units of data, such as bits, characters, blocks, or frames, transmitted during a significant measurement time interval divided by (b) the measurement time interval. *Note 2:* The effective transmission rate is usually expressed as a number of units of data per unit time, such as bits per second or characters per second. *Synonyms* average rate of transmission, effective speed of transmission.

efficiency factor. In data communications, the ratio of (a) the time to transmit a text automatically at a specified modulation rate to (b) the time actually required to receive the same text at a specified maximum error rate. *Note 1:* All of the communication facilities are assumed to be in the normal condition of adjustment and operation. *Note 2:* Telegraph communications may have different temporal efficiency factors for the two directions of transmission. *Note 3:* The practical conditions of measurement should be specified, especially the duration of the measurement.

EHF. *Abbreviation for* extremely high frequency.

EIA. *Abbreviation for* Electronic Industries Association.

EIA Class IVa fiber. *Synonym* dispersion-unshifted fiber.

EIA Class IVb fiber. *Synonym* dispersion-shifted fiber.

EIA interface. Any of a number of equipment interfaces compliant with voluntary industry standards developed by the Electronic Industries Association

(EIA) to define interface parameters. *Note 1:* Some of the EIA interface standards have been adopted by the Federal Government as Federal Standards or Federal Information Processing Standards. *Note 2:* The telecommunication-standards-developing bodies of the EIA are now part of the Telecommunications Industry Association (TIA), and the standards are designated TIA/EIA-XXX.

eight-hundred (800) service. A service that allows call originators to place toll telephone calls to 800-service subscribers, from within specified rate areas, without a charge to the call originator.

elastic buffer. 1. A buffer that has an adjustable capacity for data. 2. A buffer that introduces an adjustable delay of signals.

E layer. *See* E region, ionosphere.

electrical length. 1. Of a transmission medium, its length expressed as a multiple or submultiple of the wavelength of a periodic electromagnetic or electrical signal propagating within the medium. *Note 1:* The wavelength may be expressed in radians or in artificial units of angular measure, such as degrees. *Note 2*: In both coaxial cables and optical fibers, the velocity of propagation is approximately two-thirds that of free space. Consequently, the wavelength will be approximately two-thirds that in free space, and the electrical length, approximately 1.5 times the physical length. 2. Of a transmission medium, its physical length multiplied by the ratio of (a) the propagation time of an electrical or electromagnetic signal through the medium to (b) the propagation time of an electromagnetic wave in free space over a distance equal to the physical length of the medium in question. *Note:* The electrical length of a physical medium will always be greater than its physical length. For example, in coaxial cables, distributed resistances, capacitances and inductances impede the propagation of the signal. In an optical fiber, interaction of the lightwave with the materials of which the fiber is made, and fiber geometry, affect the velocity of propagation of the signal. 3. Of an antenna, the effective length of an element, usually expressed in wavelengths. *Note 1:* The electrical length is in general different from the physical length. *Note 2:* By the addition of an appropriate reactive element (capacitive or inductive), the electrical length may be made significantly shorter or longer than the physical length.

electrically powered telephone. A telephone in which the operating power is obtained either from a battery located at the telephone, *i.e.,* a local battery, or from a telephone central office, *i.e.,* a common battery.

electric field. The effect produced by the existence of an electric charge, such as an electron, ion, or proton, in the volume of space or medium that surrounds it. *Note:* Each of a distribution of charges contributes to the whole field at a point on the basis of superposition. A charge placed in the volume of space or in the surrounding medium has a force exerted on it.

electrochemical recording. Facsimile recording by means of a chemical reaction brought about by the passage of a signal-controlled current through the sensitized portion of the record sheet.

electrographic recording. *See* electrostatic recording.

electroluminescence. Nonthermal conversion of electrical energy into light. *Note 1:* Electroluminescence is distinguished from incandescence, which is a thermal process. *Note 2:* One example of electroluminescence is the photon emission resulting from electron-hole recombination in a pn junction, as in a light-emitting diode (LED).

electrolytic recording. Electrochemical facsimile recording in which the recorded copy is made by the passage of a signal-controlled current through an electrolyte which causes metallic ions to be deposited, thus forming an image of the object.

electromagnetic compatibility (EMC). 1. Electromagnetic compatibility is the condition which prevails when telecommunications equipment is performing its individually designed function in a common electromagnetic environment without causing or suffering unacceptable degradation due to unintentional electromagnetic interference to or from other equipment in the same environment. 2. The ability of systems, equipment, and devices that utilize the electromagnetic spectrum to operate in their intended operational environments without suffering unacceptable degradation or causing unintentional degradation because of electromagnetic radiation or response. It involves the application of sound electromagnetic spectrum management; system, equipment, and device design configuration that ensures interference-free operation; and clear concepts and doctrines that maximize operational effectiveness.

electromagnetic emission control. The control of friendly electromagnetic emissions, such as radio and radar transmissions, for the purpose of preventing or minimizing their use by unintended recipients.

electromagnetic environment (EME). 1. For a telecommunications system, the spatial distribution of electromagnetic fields surrounding a given site. *Note:* The electromagnetic environment may be expressed in terms of the spatial and temporal distribution of electric field strength (volts/meter), irradiance (watts/meter^2), or energy density (joules/meter^3). 2. The resulting product of the power and time distribution, in various frequency ranges, of the radiated or conducted electromagnetic emission levels that may be encountered by a military force, system, or platform when performing its assigned mission in its intended operational environment. It is the sum of electromagnetic interference; electromagnetic pulse; hazards of electromagnetic radiation to personnel, ordnance, and volatile materials; and natural phenomena effects of lightning and p-static.

electromagnetic interference (EMI). Any electromagnetic disturbance that interrupts, obstructs, or otherwise degrades or limits the effective performance of electronics/electrical equipment. It can be induced intentionally, as in some forms of electronic warfare, or unintentionally, as a result of spurious emissions and responses, intermodulation products, and the like. *Synonym* radio frequency interference.

electromagnetic interference (EMI) control. The control of radiated and conducted energy such that emissions that are unnecessary for system, subsystem, or equipment operation are reduced, minimized, or eliminated. *Note:* Electromagnetic radiated and conducted emissions are controlled regardless of their origin within the system, subsystem, or equipment. Successful EMI control with effective susceptibility control leads to electromagnetic compatibility.

electromagnetic intrusion. The intentional insertion of electromagnetic energy into transmission paths in any manner, with the objective of deceiving operators or of causing confusion.

electromagnetic pulse (EMP). 1. The electromagnetic radiation from a nuclear explosion caused by Compton-recoil electrons and photoelectrons from photons scattered in the materials of the nuclear device or in a surrounding medium. The resulting electric and magnetic fields may couple with electrical/electronic systems to produce damaging current and voltage surges. May also be caused by nonnuclear means. 2. A broadband, high-intensity, short-duration burst of electromagnetic energy. *Note:* In the case of a nuclear detonation, the electromagnetic pulse consists of a continuous frequency spectrum. Most of the energy is distributed throughout the lower frequencies between 3 Hz and 30 kHz.

electromagnetic radiation (EMR). Radiation made up of oscillating electric and magnetic fields and propagated with the speed of light. Includes gamma radiation, X-rays, ultraviolet, visible, and infrared radiation, and radar and radio waves.

electromagnetic radiation hazards (RADHAZ *or* EMR hazards). Hazards caused by a transmitter/antenna installation that generates electromagnetic radiation in the vicinity of ordnance, personnel, or fueling operations in excess of established safe levels or increases the existing levels to a hazardous level; or a personnel, fueling, or ordnance installation located in an area that is illuminated by electromagnetic radiation at a level that is hazardous to the planned operations or occupancy. These hazards will exist when an electromagnetic field of sufficient intensity is generated to: (a) induce or otherwise couple currents and/or voltages of magnitudes large enough to initiate electroexplosive devices or other sensitive explosive components of weapon systems, ordnance, or explosive devices; (b) cause harmful or injurious effects to humans and wildlife; (c) create sparks having sufficient magnitude to ignite flammable mixtures of materials that must be handled in the affected area.

electromagnetic spectrum. The range of frequencies of electromagnetic radiation from zero to infinity. *Note:* The electromagnetic spectrum was, by custom and practice, formerly divided into 26 alphabetically designated bands. This usage still prevails to some degree. However, the ITU formally recognizes 12 bands, from 30 Hz to 3000 GHz. New bands, from 3 THz to 3000 THz, are under active consideration for recognition. *Refer to the figure on page E-20.*

electromagnetic survivability. The ability of a system, subsystem, or equipment to resume functioning without evidence of degradation following temporary exposure to an adverse electromagnetic environment. *Note:* The system, subsystem, or equipment performance may be degraded during exposure to the adverse electromagnetic environment, but the system will not experience permanent damage, such as component burnout, that will prevent proper operation when the adverse electromagnetic environment is removed.

electromagnetic vulnerability (EMV). The characteristics of a system that cause it to suffer a definite degradation (incapability to perform the designated mission) as a result of having been subjected to a certain level of electromagnetic environmental effects.

electromagnetic wave (EMW). A wave produced by the interaction of time-varying electric and magnetic fields.

electromechanical recording. Recording by means of a signal-actuated mechanical device.

electronically controlled coupling (ECC). The coupling of a lightwave from one dielectric waveguide into another dielectric waveguide upon the application of an electric field or electrical signal. *Note:* Devices that perform ECC can be used as switches.

electronic classroom. *Synonym* teletraining.

electronic commerce. Business transactions conducted by electronic means other than conventional telephone service, *e.g.*, facsimile or electronic mail (E-mail).

electronic counter-countermeasures (ECCM). That division of electronic warfare involving actions taken to ensure friendly effective use of the electromagnetic spectrum despite the enemy's use of electronic warfare.

electronic countermeasures (ECM). That division of electronic warfare involving actions taken to prevent or reduce an enemy's effective use of the electromagnetic spectrum.

electronic deception. 1. The deliberate radiation, reradiation, alteration, suppression, absorption, denial, enhancement, or reflection of electromagnetic energy in a manner intended to convey misleading information and to deny valid information to an enemy or to enemy electronics-dependent weapons. *Note:* Among the types of electronic deception are: (a) manipulative electronic deception—Actions to eliminate revealing or convey misleading, telltale indicators that may be used by hostile forces; (b) simulative electronic deception—Actions to represent friendly notional or actual capabilities to mislead hostile forces; (c) imitative electronic deception—The introduction of electromagnetic energy into enemy systems that imitates enemy emissions. 2. Deliberate activity designed to mislead an enemy in the interpretation or use of information received by his electronic systems.

electronic emission security. Those measures taken to protect all transmissions from interception and electronic analysis.

electronic jamming. The deliberate radiation, reradiation, or reflection of electromagnetic energy for the purpose of disrupting enemy use of electronic devices, equipment, or systems.

electronic line of sight. The path traversed by electromagnetic waves that is not subject to reflection or refraction by the atmosphere.

electronic line scanning. In facsimile, a method of scanning that provides motion of the scanning spot along the scanning line by electronic means.

electronic mail (E-mail). An electronic means for communication in which (a) usually text is transmitted, (b) operations include sending, storing, processing, and receiving information, (c) users are allowed to communicate under specified conditions, and (d) messages are held in storage until called for by the addressee.

electronic message system (EMS). A message system incorporating electronic mail to a central facility which then assumes responsibility for delivering the message in hard copy form. *Note:* In DOD, these messages have a specific format known as message text format (MTF).

electronic reconnaissance. The detection, identification, evaluation, and location of foreign electromagnetic radiations emanating from other than nuclear detonations or radioactive sources.

electronics intelligence (ELINT). Technical and geolocation intelligence information derived from foreign noncommunications electromagnetic radiations emanating from other than nuclear detonations or radioactive sources.

electronics security (ELSEC). The protection resulting from all measures designed to deny unauthorized persons information of value that might be derived from their interception, and study of noncommunications electromagnetic radiations, *e.g.*, radar.

electronic switching system (ESS). 1. A telephone switching system based on the principles of time-division multiplexing of digitized analog signals. *Note:* An electronic switching system digitizes analog signals from subscribers' loops, and interconnects them by assigning the digitized signals to the appropriate time slots. It may also interconnect digital data or voice circuits. 2. A switching system with major devices constructed of semiconductor components. *Note:* A semi-electronic switching system that has reed relays or crossbar matrices, as well as semiconductor components, is also considered to be an ESS.

electronic warfare (EW). Any military action involving the use of electromagnetic and directed energy to control the electromagnetic spectrum or to attack the enemy. The three major subdivisions within electronic warfare are: electronic attack, electronic protection, and electronic warfare support.

electronic warfare support measures (ESM). 1. That division of electronic warfare involving actions taken under direct control of an operational commander to search for, intercept, identify, and locate sources of radiated electromagnetic energy for the purpose of immediate threat recognition. Thus, electronic warfare support measures (ESM) provide a source of information required for immediate decisions involving electronic countermeasures (ECM), electronic counter-countermeasures (ECCM), avoidance, targeting, and other tactical employment of forces. Electronic warfare support measures data can be used to produce signals intelligence (SIGINT), both communications intelligence (COMINT) and electronics intelligence (ELINT). 2. That division of electronic warfare involving action taken to search for, intercept, identify, and locate radiated electromagnetic energy for the purpose of immediate threat recognition. It provides a source of information required for immediate decisions involving electronic countermeasures, electronic counter-countermeasures, and other tactical actions such as avoidance, targeting and homing.

electro-optical intelligence (ELECTRO-OPTINT). Intelligence information other than signals intelligence derived from the optical monitoring of the electromagnetic spectrum from ultraviolet (0.01 μm) through the far infrared (1000 μm).

electro-optic detector. *Deprecated term. See* optoelectronic.

electro-optic effect. Any one of a number of phenomena that occur when an electromagnetic wave in the optical spectrum interacts with an electric field, or with matter under the influence of an electric field. *Note 1:* Two of the most important electro-optic effects having application as modulation mechanisms in optical communication are the Kerr effect and the Pockels effect, in which birefringence is induced or modified in a liquid (Kerr effect) or solid (Pockels effect). *Note 2:* The term *"electro-optic"* is often erroneously used as a synonym for *"optoelectronic."*

electro-optic modulator. An optical device in which a signal-controlled element is used to modulate a beam of light. *Note 1:* The modulation may be imposed on the phase, frequency, amplitude, or direction of the modulated beam. *Note 2:* Modulation bandwidths into the gigahertz range are possible using laser-controlled modulators.

electro-optics. The technology associated with those components, devices and systems which are designed to interact between the electromagnetic (optical) and the electric (electronic) state. *Note 1:* The operation of electro-optic devices depends on modification of the refractive index of a material by electric fields. *Note 2:* In a Kerr cell, the refractive index change is proportional to the square of the electric field, and the material is usually a liquid. *Note 3:* In a Pockels cell, the refractive index change varies linearly with the electric field, and the material is a crystal. *Note 4: "Electro-optic"* is often erroneously used as a synonym for *"optoelectronic"*.

ELECTRO-OPTINT. *Acronym for* electro-optical intelligence.

electrophotographic recording. Recording in which light is used to produce a change in electrostatic charge distribution to form a photographic image. *Note:* Subsequent processing is usually required to make the image visible.

electrosensitive recording. Recording in which an electrical signal is directly impressed on the record medium.

electrostatic recording. Recording by means of a signal-controlled electrostatic field. *Note:* Subsequent processing is usually required to make the image visible.

electrothermal recording. That type of recording produced principally by signal-controlled thermal action.

elemental area. In facsimile transmission systems, any segment of a scanning line, the dimension of which along the line is exactly equal to the nominal line width. *Note:* An elemental area is not necessarily the same as the scanning spot.

elementary signaling element. *See* unit interval.

elevated duct. An atmospheric duct consisting of a high-density air layer that starts at high altitudes and continues upward or remains at high altitudes, thus affecting primarily very-high-frequency (VHF) transmission.

ELF. *Abbreviation for* extremely low frequency.

ELINT. *Acronym for* electronics intelligence.

elliptical polarization. In electromagnetic wave propagation, polarization such that the tip of the electric field vector describes an ellipse in any fixed plane intersecting, and normal to, the direction of propagation. *Note 1:* An elliptically polarized wave may be resolved into two linearly polarized waves in phase quadrature with their polarization planes at right angles to each other. *Note 2:* Circular and linear polarization are special cases of elliptical polarization.

ELSEC. *Acronym for* electronics security.

E-mail. *Abbreviation for* electronic mail.

emanations security (EMSEC). The protection resulting from all measures designed to deny unauthorized persons information of value that might be derived from intercept and analysis of compromising emanations from other than crypto-equipment and telecommunications systems.

embedded base equipment. Customer-premises equipment that had been provided by the Bell Operating Companies prior to January 1, 1984, that was ordered transferred from the BOCs to AT&T by court order.

embedded customer-premises equipment. Telephone-company-provided premises equipment in use or in inventory of a regulated telephone utility as of divestiture (December 31, 1981).

embedded processor. In non-ADP equipment, a CPU and firmware that are critical to the operation of the equipment. *Note:* An embedded processor is not subject to FIRMR regulation when used for control of devices such as weapons systems, communications devices, home appliances, automobile diagnostics, *etc.*

EMC. *Abbreviation for* **electromagnetic compatibility.**

EMC analysis. Analysis of a system, subsystem, facility, or equipment to determine its electromagnetic compatibility (EMC) status. *Note:* The EMC analysis may be theoretical analysis before construction or an empirical analysis after construction.

EMCON. *Abbreviation for* emission control.

EMD. *Abbreviation for* equilibrium mode distribution.

EME. *Abbreviation for* electromagnetic environment.

Emergency Broadcast System (EBS). The EBS is composed of AM, FM, and TV broadcast stations; low-power TV stations; and non-Government industry entities operating on a voluntary, organized basis during emergencies at national, state, or operational (local) area levels.

emergency locator transmitter (ELT). A transmitter of an aircraft or survival craft actuated manually or automatically that is used as an alerting and locating aid for survival purposes.

emergency position-indicating radiobeacon station. A station in the mobile service the emissions of which are intended to facilitate search and rescue operations.

EMI. *Abbreviation for* electromagnetic interference.

emission. 1. Electromagnetic energy propagated from a source by radiation or conduction. *Note:* The emission may be either desired or undesired and may occur anywhere in the electromagnetic spectrum. 2. Radiation produced, or the production of radiation, by a radio transmitting station. For example, the energy radiated by the local oscillator of a radio receiver would not be an emission but a radiation.

emission control (EMCON). The selective and controlled use of electromagnetic, acoustic, or other emitters to optimize command and control capabilities while minimizing, for operations security (OPSEC): (a) detection by enemy sensors; (b) to minimize mutual interference among friendly systems; and/or (c) to execute a military deception plan.

emission security. Protection resulting from all measures taken to deny unauthorized persons information of value which might be derived from intercept and analysis of compromising emanations from crypto-equipment, AIS, and telecommunications systems.

emission spectrum. Of a radio emission, the distribution of power or energy as a function of frequency.

emissivity. The ratio of power radiated by a substance to the power radiated by a blackbody at the same temperature.

EMP. *Abbreviation for* electromagnetic pulse.

emphasis. In FM transmission, the intentional alteration of the amplitude-vs.-frequency characteristics of the signal to reduce adverse effects of noise in a communication system. *Note:* The high-frequency signal components are emphasized to produce a more equal modulation index for the transmitted frequency spectrum, and therefore a better signal-to-noise ratio for the entire frequency range.

EMR. *Abbreviation for* electromagnetic radiation.

EMR hazards. *Abbreviation for* electromagnetic radiation hazards.

e.m.r.p.. *Abbreviation for* effective monopole radiated power.

EMS. *Abbreviation for* electronic message system.

EMSEC. *Acronym for* emanations security.

emulate. To duplicate the functions of one system with a different system, so that the second system appears to behave like the first system. *Note:* For example, a computer emulates another, different computer by accepting the same data, executing the same programs, and achieving the same results. *Contrast with* simulate.

EMV. *Abbreviation for* electromagnetic vulnerability.

enabling signal. A signal that permits the occurrence of an event.

en-bloc signaling. Signaling in which address digits are transmitted in one or more blocks, each block containing sufficient address information to enable switching centers to carry out progressive onward routing.

encapsulation. In open systems, the technique used by layered protocols in which a lower layer protocol accepts a message from a higher layer protocol and places it in the data portion of a frame in the lower layer.

encipher. [To] Convert plain text into an unintelligible form by means of a cipher.

encode. 1. To convert data by the use of a code, frequently one consisting of binary numbers, in such a manner that reconversion to the original form is possible. 2. [To] Convert plain text to equivalent cipher text by means of a code. 3. To append redundant check symbols to a message for the purpose of generating an error detection and correction code.

encoder. *See* analog-to-digital converter.

encoding. *See* analog encoding.

encoding law. A law defining the relative values of the quantum steps used in quantizing and encoding signals.

encrypt. 1. [A] generic term encompassing encipher and encode. 2. To convert plain text into unintelligible forms by means of a cryptosystem. *Note:* The term *"encrypt"* covers the meanings of *"encipher"* and *"encode."*

end distortion. In start-stop teletypewriter operation, the shifting of the end of all marking pulses, except the stop pulse, from their proper positions in relation to the beginning of the next start pulse. *Note 1:* Shifting of the end of the stop pulse is a deviation in character time and rate rather than an end distortion. *Note 2:* Spacing end distortion is the termination of marking pulses before the proper time. *Note 3:* Marking end distortion is the continuation of marking pulses past the proper time. *Note 4:* The magnitude of the distortion is expressed as a percentage of an ideal pulse length.

end exchange. *Synonym* end office.

end finish. For an optical fiber, the optical quality of the surface at the end of the fiber.

end instrument. A communication device that is connected to the terminals of a circuit.

end office (EO). A central office at which user lines and trunks are interconnected. *Synonym* end exchange.

end-of-medium character. A control character that may be used to identify either the physical end of a data medium or the end of the usable or used portion of a data medium.

end-of-message function. In tape relay procedure, the letter and key functions, including the end-of-message indicator, that constitute the last format line.

end-of-selection character. The character that indicates the end of the selection signal.

end-of-text character (ETX). A transmission control character used to terminate text.

end-of-transmission-block character (ETB). A transmission control character used to indicate the end of a transmission block of data when data are divided into such blocks for transmission purposes.

end-of-transmission character (EOT). A transmission control character used to indicate the conclusion of a transmission that may have included one or more texts and any associated message headings. *Note:* An EOT is often used to initiate other functions, such as releasing circuits, disconnecting terminals, or placing receive terminals in a standby condition.

endpoint node. In network topology, a node connected to one and only one branch. *Synonym* peripheral node.

end system (ES). A system containing the application processes that are the ultimate source and sink of user traffic. *Note:* The functions of an end system can be distributed among two or more processors or computers.

end-to-end encryption. The encryption of information at its origin and decryption at its intended destination without any intermediate decryption.

end-to-end security. Safeguarding information in a secure telecommunication system by cryptographic or protected distribution system means from point of origin to point of destination.

endurability. The property of a system, subsystem, equipment, or process that enables it to continue to function within specified performance limits for an extended period of time, usually months, despite a severe natural or man-made disturbance, such as a nuclear attack, or a loss of external logistic or utility support. *Note:* Endurability is not compromised by temporary failures when the local capability exists to restore and maintain the system, subsystem, equipment, or process to an acceptable performance level.

endurable operation. *See* endurability.

end user. The ultimate user of a telecommunications service.

engineering channel. *Synonym* orderwire circuit.

engineering orderwire (EOW). *Synonym* orderwire circuit.

enhanced-quality television. *Synonym [in CCITT usage]* improved-definition television.

enhanced service. Service, offered over commercial carrier transmission facilities used in interstate communications, that employs computer processing applications that act on the format, content, code, protocol, or similar aspects of the subscriber's transmitted information; provides the subscriber with additional, different, or restructured information; or involves subscriber interaction with stored information.

ENQ. *Abbreviation for* enquiry character.

enquiry character (ENQ). A transmission control character used as a request for a response from the station with which a connection has been set up. *Note:* The response may include station identification, the type of equipment in service, and the status of the remote station.

***E/N* ratio.** In the transmission of a pulse of an electromagnetic wave representing a bit, the ratio of

(a) the energy in each bit, *E*, to (b) the noise energy density per hertz, *N*. *Note: E* is usually expressed in joules per bit and *N* is usually expressed in watts per hertz. Thus, the *E/N* ratio is hertz-seconds per bit. A joule is a watt-second and a hertz is a cycle per second. Thus, the *E/N* ratio is actually cycles per bit. However, if a cycle is a bit, then the *E/N* ratio is dimensionless.

entrance facility. The entrance to a building for both public and private network service cables (including antenna transmission lines, where applicable), including the entrance point at the building wall or floor, and continuing to the entrance room or entrance space.

entrance point. In a building, the point of emergence of telecommunications service cables through an exterior wall, floor slab, or from a rigid metal conduit or intermediate metal conduit.

entrance room. In a building, a space in which the joining of inter- and/or intrabuilding telecommunications backbone facilities takes place. *Note:* An entrance room may serve also as an equipment room.

envelope. The boundary of the family of curves obtained by varying a parameter of a wave.

envelope delay distortion. Signal distortion that results when the rate of change of phase shift with frequency over the necessary bandwidth of the signal is not constant. *Note:* Envelope delay distortion is usually expressed as one-half the difference between the delays of the two extremes of the necessary bandwidth.

environmental control. *See* air-conditioning.

environmental security. 1. The security that is inherent in the physical surroundings in which a facility or functional unit is located, such as on ships, on aircraft, and in underground vaults, where locations by their nature provide a certain amount of protection against exploitation of compromising emanation even before other protective measures are implemented. 2. The application of electrical, acoustic, physical, and other safeguards to an area to minimize the risk of unauthorized interception of information from the area.

EO. *Abbreviation for* end office.

EOT. *Abbreviation for* end-of-transmission character.

EOW. *Abbreviation for* engineering orderwire.

E-plane bend. *Synonym* E-bend.

epoch date. A date in history, chosen as the reference date from which time is measured. *Note 1:* An example of an epoch date is the beginning instant of January 1, 1900, Universal Time, for Transmission Control Protocol/Internet Protocol (TCP/IP). *Note 2:* TCP/IP programs exchange date or time-of-day information with time expressed as the number of seconds past the epoch date.

equal gain combiner. A diversity combiner in which the signals on each channel are added. *Note:* The channel gains can be made to remain always and everywhere equal so that the resultant signal remains approximately constant.

equalization. The maintenance of system transfer function characteristics within specified limits by modifying circuit parameters. *Note:* Equalization includes modification of circuit parameters, such as resistance, inductance, or capacitance.

equal-length code. A telegraph or a data code in which (a) all the words or code groups are composed of the same number of unit elements, (b) each element has the same duration or spatial length, (c) each word or code group has the same duration or spatial length, and (d) usually each word or code group has the same number of characters.

equal-level patch bay. An analog patching facility at which all nominal input and output voice frequency levels are uniform. *Note:* The use of an equal-level patch bay permits patching without making transmission level adjustments.

equatorial orbit. For a satellite orbiting the Earth, an orbit in the equatorial plane. *Note:* An equatorial orbit has an inclination angle of 0°.

equilibrium coupling length. *Synonym* equilibrium length.

equilibrium length. For a specific excitation condition, the length of multimode optical fiber necessary to attain equilibrium mode distribution. *Note:* Equilibrium length is sometimes used to refer to the longest such length, as would result from a worst-case, but undefined, excitation. *Synonyms* equilibrium coupling length, equilibrium mode distribution length.

equilibrium mode distribution (EMD). That condition in a multimode fiber wherein after propagation has taken place for a certain distance, called the *"equilibrium length,"* the relative power distribution among modes becomes statistically constant and remains so for the course of further propagation down the fiber. *Note 1:* In practice, the equilibrium length may vary from a fraction of a kilometer to more than a kilometer. *Note 2:* After the equilibrium length has been traversed, the numerical aperture of the fiber's output is independent of the numerical aperture of the optical source, *i.e.*, beam, that drives the fiber. This is because of mode coupling and stripping, primarily by small perturbations in the fiber's geometry which result from the manufacturing and cabling processes. *Note 3:* In the ray-optics analogy, the equilibrium mode distribution may be loosely thought of as a condition in which the "outermost rays" in the fiber core are stripped off by such phenomena as microbends, and only the "innermost rays" continue to propagate. In a typical 50-μm core multimode graded-index fiber, light propagating under equilibrium conditions occupies essentially the middle seven-tenths of the core and has a numerical aperture approximately seven-tenths that of the full numerical aperture of the fiber. This is why in-line optical attenuators based on the principle of gap loss may be ineffective or induce a lower-than-rated loss if they are inserted near the optical receiver. To be fully effective, gap-loss attenuators should be inserted near the optical transmitter, where the core is fully filled. *Synonyms* equilibrium mode power distribution, steady-state condition.

equilibrium mode distribution length. *Synonym* equilibrium length.

equilibrium mode power distribution. *Synonym* equilibrium mode distribution.

equilibrium mode simulator. For an optical fiber, a device or optical system used to create an approximation of the equilibrium mode distribution.

equipment clock. A clock that satisfies the particular needs of equipment and, in some cases, may control the flow of data at the equipment interface.

equipment intermodulation noise. Intermodulation noise introduced into a system by a specific piece of equipment.

equipment room. In a building, a centralized space for telecommunications equipment that serves the occupants of the building. *Note:* An equipment room is considered distinct from a telecommunications closet because of the nature or complexity of the equipment housed by the equipment room.

equipment side. The portion of a device that is directly connected to facilities internal to a station, such as the data terminal equipment (DTE) side of the DTE/ data-circuit-terminating (DCE) interface, switches, and user end instruments.

equipotential ground plane. A mass, or bonded masses, of conducting material that offer a negligible impedance to current flow. *Note:* Equipotential ground planes may be in direct contact with the earth or may be physically isolated from the earth and suitably connected to it.

equivalent network. 1. In a system, a network that may replace another network without altering the performance of the system. 2. A network with external characteristics that are identical to those of another network. 3. A theoretical representation of an actual network.

equivalent noise resistance. A quantitative representation in resistance units of the spectral density of a noise-voltage generator, given by $R_n = (\pi W_n)/(kT_0)$, where W_n is the spectral density, k is Boltzmann=s constant, T_0 is the standard noise temperature (290 K), and $kT_0 = 4.00$ H 10^{B21} watt-seconds. *Note:* The equivalent noise resistance in terms of the mean-square noise-generator voltage, e^2, within a frequency increment, Δf, is given by $R_n = e^2/(4kT_0\Delta f)$.

equivalent noise temperature. The temperature, usually expressed in kelvins, of a hypothetical matched resistance at the input of an assumed noiseless device, such as a noiseless amplifier, that would account for the measured output noise.

equivalent pulse code modulation (PCM) noise. The amount of thermal noise power on a frequency-division multiplexed (FDM) or wire channel necessary to approximate the same judgment of speech quality created by quantizing noise in a PCM channel. *Note 1:* The speech quality judgment is based on comparative tests. *Note 2:* Generally, 33.5 dBrnC ± 2.5 dB is considered the approximate equivalent PCM noise of a 7-bit PCM system.

equivalent satellite link noise temperature. The noise temperature referred to the output of the receiving antenna of the Earth station corresponding to the radio-frequency noise power which produces the total observed noise at the output of the satellite link excluding noise due to interference coming from satellite links using other satellites and from terrestrial systems.

erase. 1. To obliterate information from a storage medium, such as to clear or to overwrite. 2. In a magnetic storage medium, to remove all stored data by (a) changing the medium to an unmagnetized state or (b) changing the medium to a predetermined magnetized state. 3. In paper tape and punched card storage, to punch a hole at every punch position.

erect position. In frequency-division multiplexing, a position of a translated channel in which an increase signal frequency in the untranslated channel causes an increase signal frequency in the translated channel. *Synonym* **upright position.**

E region. That portion of the ionosphere existing between approximately 95 and 130 km above the surface of the Earth. *Note:* The E Region lies between the D and F regions. *Synonyms* Heaviside layer, Kennelly-Heaviside layer.

erlang. A dimensionless unit of the average traffic intensity (occupancy) of a facility during a period of time, usually a busy hour. *Note 1:* Erlangs, a number between 0 and 1, inclusive, is expressed as the ratio of (a) the time during which a facility is continuously or cumulatively occupied to (b) the time that the facility is available for occupancy. *Note 2:* Communications traffic, measured in erlangs for a period of time, and offered to a group of shared facilities, such as a trunk group, is equal to the average of the traffic intensity, in erlangs for the same period of time, of all individual sources, such as telephones, that share and are served exclusively by this group of facilities. *Synonym* traffic unit.

erroneous block. A block in which there are one or more erroneous bits.

error. 1. The difference between a computed, estimated, or measured value and the true, specified, or theoretically correct value. 2. A deviation from a correct value caused by a malfunction in a system or a functional unit. *Note:* An example of an error is the occurrence of a wrong bit caused by an equipment malfunction.

error blocks. In video systems, a form of block distortion in which a block or blocks in the received image bear no resemblance to the current or previous scene and may contrast greatly with adjacent blocks.

error budget. The bit-error-ratio requirement allocated to the respective segments of a communications system, such as trunking, switching, access, and terminal devices, in a manner that satisfies the specified system end-to-end bit-error-ratio requirement for transmitted traffic.

error burst. A contiguous sequence of symbols, received over a data transmission channel, such that the first and last symbols are in error and there exists no contiguous subsequence of *m* correctly received symbols within the error burst. *Note:* The integer parameter *m* is referred to as the guard band of the error burst. The last symbol in a burst and the first symbol in the following burst are accordingly separated by *m* correct bits or more. The parameter *m* should be specified when describing an error burst.

error control. Any technique that will detect or correct errors.

error-correcting code. A code in which each telegraph or data signal conforms to specific rules of construction so that departures from this construction in the received signal can generally be automatically detected and corrected. *Note 1:* If the number of errors is less than or equal to the maximum correctable threshold of the code, all errors will be corrected. *Note 2:* Error-correcting codes require more signal elements than are necessary to convey the basic information. *Note 3:* The two main classes of error-correcting codes are block codes and convolutional codes.

error-correcting system. In digital data transmission, a system employing either forward error correction (FEC) or automatic repeat-request (ARQ) techniques such that most transmission errors are automatically removed from the data unit prior to delivery to the destination facility.

error-detecting-and-feedback system. *Synonym* ARQ.

error-detecting code. A code in which each telegraph or data signal conforms to specific rules of construction, so that departures from this construction in the received signal can generally be detected automatically. *Note:* Error-detecting codes require more signal elements than are necessary to convey the basic information.

error-detecting system. A system employing an error-detecting code and so arranged that any signal detected as being in error is either deleted from the data delivered to the data sink, in some cases with an indication that such deletion has taken place, or delivered to the data sink together with an indication that the signal is in error.

error message. In a computer or communications system, a message that indicates that an error has been made and, sometimes, the nature or type of error.

error rate. *Deprecated term. See* error ratio.

error ratio. The ratio of the number of bits, elements, characters, or blocks incorrectly received to the total number of bits, elements, characters, or blocks sent during a specified time interval. *Note:* For a given communication system, the bit error ratio will be affected by both the data transmission rate and the signal power margin.

error signal. In computer and communications systems, an audio or visual signal that indicates that an error has been made by the system or its operator. *Note:* In most systems, the error signal accompanies an error message and is used to draw operator attention to the error message.

e.r.p. [*or* ERP]. *Abbreviation for* effective radiated power.

ES. *Abbreviation for* end system, expert system.

ESC. *Abbreviation for* escape character.

escape character (ESC). 1. In alphabet coding schemes, a specially designated character, the occurrence of which in the data signifies that one or more of the characters to follow are from a different character code, *i.e.,* have meanings other than normal. 2. In a text-control sequence of characters, a control character that indicates the beginning of the sequence and the end of any preceding text.

ESF. *Abbreviation for* extended superframe.

ESM. *Abbreviation for* electronic warfare support measure.

ESS. *Abbreviation for* electronic switching system.

essential service. A network-provided service feature in which a priority dial tone is furnished. *Note 1:* Essential service is typically provided to fewer than 10 % of network users. *Note 2:* Essential service is recommended for use in conjunction with NS/EP telecommunications services. *Synonym* critical service.

ETB. *Abbreviation for* end-of-transmission-block character.

Ethernet. A standard protocol (IEEE 802.3) for a 10-Mb/s baseband local area network (LAN) bus using carrier-sense multiple access with collision detection (CSMA/CD) as the access method, implemented at the Physical Layer in the ISO Open Systems Interconnection–Reference Model, establishing the physical characteristics of a CSMA/CD network. *Note 1:* Ethernet is a standard for using various transmission media, such as coaxial cables, unshielded twisted pairs, and optical fibers. *Note 2:* The IEEE-802.3 standard is based on a proprietary product with a similar name.

ETX. *Abbreviation for* end-of-text character.

evanescent field. In a waveguide, a time-varying field having an amplitude that decreases monotonically as a function of transverse radial distance from the waveguide, but without an accompanying phase shift. *Note 1:* The evanescent field is coupled, *i.e.*, bound, to an electromagnetic wave or mode propagating inside the waveguide. *Note 2:* The evanescent field is a surface wave. *Note 3:* In fiber optics, the evanescent field may be used to provide coupling to another fiber.

evanescent mode. A mode of the evanescent field.

even parity. *See* parity, parity check.

event. 1. An occurrence or happening, usually significant to the performance of a function, operation, or task. 2. In Integrated Services Digital Networks (ISDN), an instantaneous occurrence that changes at least one of the attributes of the global status of a managed object. *Note:* An event (a) may be persistent or temporary, thus allowing for functions, such as surveillance, monitoring, and performance measurement, (b) may generate reports, (c) may be spontaneous or planned, (d) may trigger other events, and (e) may be triggered by one or more other events.

EW. *Abbreviation for* electronic warfare.

exalted-carrier reception. A method of receiving either amplitude- or phase-modulated signals in which method the carrier is separated from the sidebands, filtered and amplified, and then combined with the sidebands again at a higher level prior to demodulation. *Synonym* reconditioned carrier reception.

exception condition. In data transmission, the condition assumed by a device when it receives a command that it cannot execute.

excess insertion loss. *Deprecated term. See* insertion loss. *Note: Excess insertion loss* was used to indicate that, in an optical-fiber coupler, the loss occasioned by dividing the input power among the ports is not the total insertion loss.

exchange. 1. A room or building equipped so that telephone lines terminating there may be interconnected as required. *Note:* The equipment may include manual or automatic switching equipment. 2. In the telephone industry, a geographic area (such as a city and its environs) established by a regulated telephone company for the provision of local telephone services. 3. In the Modification of Final Judgment (MFJ), a local access and transport area.

exchange access. In telephone networks, access in which exchange services are provided for originating or terminating interexchange telecommunications within the exchange area.

exchange area. A geographic area served by one or more central offices within which local telephone service is furnished under regulation.

exchange facilities. The facilities included within a local access and transport area.

executive program. *Synonym* supervisory program.

exempted addressee. An organization, activity, or person included in the collective address group of a message and deemed by the message originator as having no need for the information in the message. *Note:* Exempted addressees may be explicitly excluded from the collective address group for the particular message to which the exemption applies.

existing quality television. *Synonym [in CCITT usage]* distribution-quality television.

expander. A device that restores the dynamic range of a compressed signal to its original dynamic range.

expansion. The restoration of the dynamic range of a compressed signal to its original dynamic range.

expansion capability. The inherent limit for increasing the capacity of a system beyond its installed capacity.

expedited data unit. In layered systems, a service data unit that is delivered to a peer entity in the destination open system before the delivery of any subsequent service data unit sent on that connection.

experimental station. A station utilizing radio waves in experiments with a view to the development of science or technique. This definition does not include amateur stations.

expert system (ES). A computer system that facilitates solving problems in a given field or application by drawing inference from a knowledge base developed from human expertise. *Note 1:* The term *"expert system"* is sometimes used synonymously with *"knowledge-based system,"* although it is usually taken to emphasize expert knowledge. *Note 2:* Some expert systems are able to improve their knowledge base and develop new inference rules based on their experience with previous problems.

express orderwire. A permanently connected voice circuit between selected stations for technical control purposes.

extended area service (EAS). A network-provided service feature in which a user pays a higher flat rate to obtain wider geographical coverage without paying per-call charges for calls within the wider area.

extended binary coded decimal interchange code. *See* EBCDIC.

extended-definition television (EDTV). Television in which (a) improvements are made to the standard National Television System Committee (NTSC) television system, (b) the improvements are receiver-compatible with the standard NTSC television system, and (c) the improvements modify the standard NTSC television system emission standards. *Note 1:* EDTV improvements may include (a) a wider aspect ratio, (b) a higher picture definition than NTSC definition, and (c) any of the improvements used in improved-definition television (IDTV). *Note 2:* When EDTV signals are transmitted in the 4:3 aspect ratio, it is referred to as *"EDTV."* When transmitted in a wider aspect ratio, it is referred to as *"EDTV-Wide."*

extended superframe. A T-carrier framing technique in which framing requiring less frequent synchronization than the original T-carrier superframe format is provided for D-4 formatting and for on-line, real-time testing of circuit capability and operating condition. *Note:* Less-frequent synchronization frees overhead bits for use in testing and monitoring.

extension bell. In telephony, a user end device, separate from a subscriber telephone, which device produces an audible signal indicating that there is an incoming call from a switchboard or exchange.

extension facility. A facility that provides access to communications for a user or group of users isolated from a central communications node.

extension terminal. A terminal that is added to an existing terminal and that uses the same circuit and address, *i.e.,* port and number, as the terminal to which it is added.

external timing reference. In a given communications system, a timing reference obtained from a source, such as a navigation system, external to the given system. *Note:* External timing references are usually referenced to Coordinated Universal Time.

extinction coefficient. The sum of the absorption coefficient and the scattering coefficient.

extinction ratio (r_e). The ratio of two optical power levels,

$$r_\theta = \frac{P_1}{P_2},$$

of a digital signal generated by an optical source, *e.g.,* a laser diode, where P_1 is the optical power level generated when the light source is "on," and P_2 is the power level generated when the light source is "off." *Note:* The extinction ratio may be expressed as a fraction or in dB.

extra bit. *Synonym* added bit.

extra block. *Synonym* added block.

extremely high frequency (EHF). Frequencies from 30 GHz to 300 GHz.

extremely low frequency (ELF). Frequencies from 30 Hz to 300 Hz.

extrinsic joint loss. For an optical fiber, that portion of a joint loss that is not intrinsic to the fibers, *e.g.*, loss caused by end separation, angular misalignment, or lateral misalignment.

eye pattern. An oscilloscope display in which a pseudorandom digital data signal from a receiver is repetitively sampled and applied to the vertical input, while the data rate is used to trigger the horizontal sweep. *Note:* System performance information can be derived by analyzing the display. An open eye pattern corresponds to minimal signal distortion. Distortion of the signal waveform due to intersymbol interference and noise appears as closure of the eye pattern.

eyes only. A message marker for a special-category message that is intended for delivery only to a specific person, or authorized representative of that person, and therefore no one else.

facet erosion. In laser diodes, a phenomenon in which a high field intensity of stimulated optical radiation causes degradation of the facets, *i.e.,* those forming the cavity mirrors, decreasing reflectivity and resulting in a decrease of the internal quantum efficiency and an increase in the threshold current.

facility. 1. A fixed, mobile, or transportable structure, including (a) all installed electrical and electronic wiring, cabling, and equipment and (b) all supporting structures, such as utility, ground network, and electrical supporting structures. 2. A network-provided service to users or the network operating administration. 3. A transmission pathway and associated equipment. 4. In a protocol applicable to a data unit, such as a block or frame, an additional item of information or a constraint encoded within the protocol to provide the required control. 5. A real property entity consisting of one or more of the following: a building, a structure, a utility system, pavement, and underlying land.

facility grounding system. The electrically interconnected system of conductors and conductive elements that (a) provides multiple current paths to the earth electrode subsystem, and (b) consists of the earth electrode subsystem, the lightning protection subsystem, and the fault protection subsystem.

facsimile (FAX). 1. A form of telegraphy for the transmission of fixed images, with or without half-tones, with a view to their reproduction in a permanent form. In this definition the term telegraphy has the same general meaning as defined in the Convention. 2. The process by which fixed graphic images, such as printed text and pictures, are scanned, and the information converted into electrical signals that may be transmitted over a telecommunications system and used to create a copy of the original, or an image so produced. *Note 1:* Wirephoto and telephoto are facsimile via wire circuits. Radiophoto is facsimile via radio. *Note 2:* Technology now exists that permits the transmission and reception of facsimile data to or from a computer without requiring hard copy at either end. *Note 3:* Current facsimile systems are designated and defined as follows:

Group 1 Facsimile. The mode of black and white facsimile operation, defined in CCITT Recommendation T.2, that uses double sideband modulation without any special measures to compress the bandwidth. *Note 1:* A 216 × 279-mm document, *i.e.*, an 8½ × 11-inch document, may be transmitted in approximately 6 minutes via a telephone-type circuit. Additional modes in this group may be designed to operate at a lower resolution suitable for the transmission of 216 × 279-mm documents in 3 to 6 minutes. *Note 2:* The CCITT frequencies used are 1300 Hz for white and 2300 Hz for black. The North American standard is 1500 Hz for white and either 2300 or 2400 Hz for black.

Group 2 Facsimile. The mode of black and white facsimile operation, defined in CCITT Recommendation T.3, that accomplishes bandwidth compression by using encoding and vestigial sideband, but excludes processing of the document signal to reduce redundancy. *Note:* A 216 × 279-mm document, *i.e.*, an 8½ × 11-inch document, may be transmitted in approximately 3 minutes using a 2100-Hz AM/PM/VSB, over a telephone-type circuit.

Group 3 Facsimile. The mode of black and white facsimile operation, defined in ITU-T Recommendation T.4, that incorporates means for reducing the redundant information in the signal by using a one-dimensional run-length coding scheme prior to the modulation process. *Note 1:* A 216 × 279-mm document, *i.e.*, an 8½ × 11-inch document, may be transmitted in approximately 1 minute or less over a telephone-type circuit with twice the Group 2 horizontal resolution. The vertical resolution may also be doubled. *Note 2:* Group 3 Facsimile machines have integral digital modems. *Note 3:* An optional two-dimensional bandwidth compression scheme is also defined within the Group 3 Facsimile Recommendation. *Note 4:* When any CCITT or CCIR Recommendation is modified by the ITU-T, the modified document is designated as an ITU-T Recommendation.

Group 3C Facsimile. The Group 3 digital mode of facsimile operation defined in CCITT Recommendation T.30. *Note:* Group 3C is also referred to as Group 3 Option C or as Group 3-64 kb/s.

Group 4 Facsimile. The mode of black and white facsimile operation defined in ITU-T Recommendation T.563 and CCITT Recommendation T.6. *Note 1:* Group 4 Facsimile uses bandwidth compression techniques to transmit, essentially without errors, a 216 × 279-mm document, *i.e.*, an 8½ × 11-inch document, at a nominal resolution of 8 lines/mm in less than 1 minute over a public data network voice-grade circuit. *Note 2:* When any CCITT or CCIR Recommendation is modified by the ITU-T, the modified document is designated as an ITU-T Recommendation.

Type I Facsimile. The mode of digital black and white facsimile operation defined in MIL-STD-188-161 used for transmission of bi-level information (*e.g.*, text and simple graphics). *Note:* Type I facsimile is interoperable with the black-and-white facsimile mode of STANAG 5000 and is designed for operation over noisy communications links such as tactical channels.

Type II Facsimile. The mode of gray-scale facsimile operation defined in MIL-STD-188-161 used for transmission of multi-level information (*e.g.*, photographs). *Note:* Type II facsimile is interoperable with the black-and-white facsimile mode of Type I or STANAG 5000 equipment and is designed for operation over noisy communications links such as tactical channels.

facsimile converter. 1. In a facsimile receiver, a device that changes the signal modulation from frequency-shift keying (FSK) to amplitude modulation (AM). 2. In a facsimile transmitter, a device that changes the signal modulation from amplitude modulation (AM) to frequency-shift keying (FSK).

facsimile frequency shift. At any point in a frequency-shift facsimile system, the numerical difference between the frequency that corresponds to a white signal and the frequency that corresponds to a black signal. *Note:* Facsimile frequency shift is usually expressed in hertz.

facsimile picture signal. In facsimile systems, the baseband signal that results from the scanning process.

facsimile receiver. In a facsimile system, the equipment that converts the facsimile picture signal into a recorded copy.

facsimile recorder. In a facsimile receiver, the device that performs the final conversion of the facsimile picture signal to an image of the object, *i.e.*, makes the recorded copy.

facsimile signal level. In a facsimile system, the signal level at any point in the system. *Note 1:* The facsimile signal level is used to establish the operating levels. *Note 2:* The facsimile signal level is usually expressed in dB with respect to some standard value, such as 1 mW (milliwatt), *i.e.*, 0 dBm.

facsimile transceiver. In a facsimile system, the equipment that sends and receives facsimile signals. *Note:* Full-duplex facsimile transceivers can send and receive at the same time; half-duplex facsimile transceivers cannot.

facsimile transmitter. In a facsimile system, the equipment that converts the baseband picture signals, *i.e.*, the baseband signals resulting from scanning the object, into signals suitable for transmission by a communications system.

fade margin. 1. A design allowance that provides for sufficient system gain or sensitivity to accommodate expected fading, for the purpose of ensuring that the required quality of service is maintained. 2. The amount by which a received signal level may be reduced without causing system performance to fall below a specified threshold value. *Synonym* fading margin.

fading. In a received signal, the variation (with time) of the amplitude or relative phase, or both, of one or more of the frequency components of the signal. *Note:* Fading is caused by changes in the characteristics of the propagation path with time.

fading distribution. The probability distribution that signal fading will exceed a given value relative to a specified reference level. *Note 1:* In the case of phase interference fading, the time distribution of the instantaneous field strength usually approximates a Rayleigh distribution when several signal components of equal amplitude are present. *Note 2:* The field strength is usually measured in volts per meter. *Note 3:* The fading distribution may also be measured in terms of power level, where the unit of measure is usually watts per square meter and the expression is in dB.

fading margin. *Synonym* fade margin.

fail. *See* failure, graceful degradation.

fail safe. 1. Of a device, the capability to fail without detriment to other devices or danger to personnel. 2. Pertaining to the automatic protection of programs and/or processing systems to maintain safety when a hardware or software failure is detected in a system. 3. Pertaining to the structuring of a system such that either it cannot fail to accomplish its assigned mission regardless of environmental factors or that the probability of such failure is extremely low.

fail-safe operation. 1. Operation that ensures that a failure of equipment, process, or system does not propagate beyond the immediate environs of the failing entity. 2. A control operation or function that prevents improper system functioning or catastrophic degradation in the event of circuit malfunction or operator error.

failure. The temporary or permanent termination of the ability of an entity to perform its required function.

failure access. [An] Unauthorized and usually inadvertent access to data resulting from a hardware or software failure in an AIS.

fair queuing. The controlling of congestion in gateways by restricting every host to an equal share of gateway bandwidth. *Note:* Fair queuing does not distinguish between small and large hosts or between hosts with few active connections and those with many.

fall time. The time required for the amplitude of a pulse to decrease (fall) from a specified value (usually 90 percent of the peak value exclusive of overshoot or undershoot) to another specified value (usually 10 percent of the peak value exclusive of overshoot or undershoot). *Note:* Limits on undershoot and oscillation, *i.e.,* hunting, may need to be specified when specifying fall time limits. *Synonym* pulse decay time.

false character. *See* illegal character.

false clock. A condition where a phase-locked loop controlling a clock locks on a frequency other than the correct frequency. *Note 1:* False clock can occur when there is excessive phase shift, as a function of frequency, in the loop. *Note 2:* False clock often occurs where the false frequency is a harmonic of the correct frequency.

false lock. A condition where a phase-locked loop locks to a frequency other than the correct one, or to an improper phase.

fan-beam antenna. A directional antenna producing a main beam having a large ratio of major to minor dimension at any transverse cross section.

fan out. *Synonym* break out.

FAQ file. *Abbreviation for* Frequently Asked Questions file. An online file that contains frequently asked questions with answers provided to assist new users and avoid repetitive offline inquiries. *Note:* An *FAQ file* is usually created for Internet news groups, but is also used in other applications.

Faraday effect. A magneto-optic effect in which the polarization plane of an electromagnetic wave is rotated under the influence of a magnetic field parallel to the direction of propagation. *Note:* The Faraday effect may be used to modulate a lightwave.

far-end crosstalk. Crosstalk that is propagated in a disturbed channel in the same direction as the propagation of a signal in the disturbing channel. *Note:* The terminals of the disturbed channel, at which the far-end crosstalk is present, and the energized terminals of the disturbing channel, are usually remote from each other.

far field. *Synonym* far-field region.

far-field diffraction pattern. The diffraction pattern of a source (such as an LED, ILD, or the output end of an optical fiber) observed at an infinite distance from the source. *Note 1:* A far-field pattern exists at distances that are large compared with s^2/λ, where s is a characteristic dimension of the source and λ is the wavelength. For example, if the source is a uniformly illuminated circle, then s is the radius of the circle. *Note 2:* The far-field diffraction pattern of a source may be observed at infinity or (except for scale) in the focal plane of a well-corrected lens. The far-field pattern of a diffracting screen illuminated by a point source may be observed in the image plane of the source. *Synonym* Fraunhofer diffraction pattern. *Contrast with* near-field diffraction pattern.

far-field radiation pattern. A radiation pattern measured at the far field of an antenna or other emitter.

far-field region. The region where the angular field distribution is essentially independent of distance from the source. *Note 1:* If the source has a maximum overall dimension D that is large compared to the wavelength, the far-field region is commonly taken to exist at distances greater than $2D^2/\lambda$ from the source, λ being the wavelength. *Note 2:* For a beam focused at infinity, the far-field region is sometimes referred to as the Fraunhofer region. *Synonyms* far field, far zone, Fraunhofer region, radiation field.

far zone. *Synonym* far-field region.

fast packet switching. A packet switching technique that increases the throughput by eliminating overhead. *Note 1:* Overhead reduction is accomplished by allocating flow control and error correction functions to either the user applications or the network nodes that interface with the user. *Note 2:* Cell relay and frame relay are two implementations of fast packet switching.

fast select. An optional user facility in the virtual call service of CCITT X.25 protocol that allows the inclusion of user data in the call request/connected and clear indication packets. *Note:* Fast select is an essential feature of the CCITT X.25 (1984) protocol.

fault. 1. An accidental condition that causes a functional unit to fail to perform its required function. 2. A defect that causes a reproducible or catastrophic malfunction. *Note:* A malfunction is considered reproducible if it occurs consistently under the same circumstances. 3. In power systems, an unintentional short-circuit, or partial short-circuit, between energized conductors or between an energized conductor and ground.

fault management. In network management, the set of functions that (a) detect, isolate, and correct malfunctions in a telecommunications network, (b) compensate for environmental changes, and (c) include maintaining and examining error logs, accepting and acting on error detection notifications, tracing and identifying faults, carrying out sequences of diagnostics tests, correcting faults, reporting error conditions, and localizing and tracing faults by examining and manipulating database information.

fault protection subsystem. In a facility power distribution system, the subsystem that provides a direct path from each power sink to the earth electrode subsystem. *Note:* The fault protection subsystem is usually referred to as a *"green wire."*

fault tolerance. The extent to which a functional unit will continue to operate at a defined performance level even though one or more of its components are malfunctioning.

FAX. *Acronym for* facsimile.

FC. *Abbreviation for* functional component.

FCC. The U.S. Government board of five presidential appointees that has the authority to regulate all non-Federal Government interstate telecommunications (including radio and television broadcasting) as well as all international communications that originate or terminate in the United States. *Note:* Similar authority for regulation of Federal Government telecommunications is vested in the National Telecommunications and Information Administration (NTIA).

FCC registration program. The Federal Communications Commission program and associated directives intended to assure that all connected terminal equipment and protective circuitry will not harm the public switched telephone network or certain private line services. *Note 1:* The FCC registration program requires the registering of terminal equipment and protective circuitry in accordance with Subpart C of part 68, Title 47 of the *Code of Federal Regulations.* This includes the assignment of identification numbers to the equipment and the testing of the equipment. *Note 2:* The FCC registration program contains no requirement that accepted terminal equipment be compatible with, or function with, the network.

FCS. *Abbreviation for* frame check sequence. *See* cyclic redundancy check.

FDDI. *Abbreviation for* fiber distributed data interface.

FDDI-2. *See* fiber distributed data interface.

FDHM. *See* full width at half maximum.

FDM. *Abbreviation for* frequency-division multiplexing.

FDMA. *Abbreviation for* frequency-division multiple access.

FDX. *Abbreviation for* full duplex.

FEC. *Abbreviation for* forward error correction.

Federal Communications Commission. *See* FCC.

Federal Telecommunications System (FTS). A switched long-distance telecommunications service formerly provided for official Federal Government use. *Note:* FTS has been replaced by Federal Telecommunications Service 2000 (FTS2000).

Federal Telecommunications System 2000 service. *See* FTS2000.

feed. 1. To supply a signal to the input of a system, subsystem, equipment, or component, such as a transmission line or antenna. 2. A coupling device between an antenna and its transmission line. *Note:* A feed may consist of a distribution network or a primary radiator. 3. A transmission facility between (a) the point of origin of a signal, such as is generated in a radio or television studio, and (b) the head-end of a distribution facility, such as a broadcasting station in a network. 4. Pertaining to the function of inserting one thing into another, such as in a feed horn, paper feed, card feed, and line feed.

feedback. 1. The return of a portion of the output, or processed portion of the output, of a (usually active) device to the input. *Note 1:* The feedback signal will have a certain magnitude and phase relationship relative to the output signal or the input signal. This relationship can be used to influence the behavior, such as the gain and stability, of the overall circuit. *Note 2:* If the feedback is regenerative (additive), it is called "positive feedback," which increases gain and distortion, and decreases linearity and stability. *Note 3:* If the feedback is degenerative (subtractive), it is called "negative feedback," which reduces the gain and distortion, and increases linearity and stability. *Note 4:* Feedback may occur inadvertently, and be detrimental. 2. Information returned as a response to an originating source.

feeder echo noise. Signal distortion resulting from reflected waves in a transmission line that is many wavelengths long and mismatched at both the generator and the load ends.

feeder link. A radio link from an Earth station at a given location to a space station, or vice versa, conveying information for a space radiocommunication service other than for the fixed-satellite service. The given location may be at a specified fixed point, or at any fixed point within specified areas.

FEP. *Abbreviation for* front-end processor.

Fermat's principle. A principle stating that a ray of light follows the path that requires the least time to travel from one point to another, including reflections and refractions that may occur. *Synonym* least-time principle.

fetch protection. [An] AIS-provided restriction to prevent a program from accessing data in another user's segment of storage.

FET photodetector. A photodetector using photo-generation of carriers in the channel region of a field-effect transistor structure to provide photodetection with current gain.

fiber. *See* optical fiber.

fiber amplifier. A device that amplifies an optical signal directly, without the need to convert it to an electrical signal, amplify it electrically, and reconvert it to an optical signal. *Note 1:* One type of fiber amplifier uses a doped fiber (*e.g.*, a fiber doped with erbium), which bears the communication signal, and which is optically pumped with a laser having a high-powered continuous output at an optical frequency slightly higher than that of the communication signal. The signal is intensified by Raman amplification. *Note 2:* Because neither optical-electrical conversion nor electrical amplification takes place, this type of amplifier is well suited for a wide variety of applications, both digital and analog. *Note 3:* Because this type of amplifier does not require extraordinary frequency (wavelength) control of the pumping laser, it is relatively simple. *Synonym* Raman amplifier.

fiber axis. The longitudinal center of symmetry of an optical fiber, *i.e.*, the locus of points that are determined by the centers of mechanical symmetry of the outside diameters of fiber cross sections sampled continuously along the length of the fiber.

fiber bandwidth. *See* bandwidth (of an optical fiber).

fiber buffer. *See* buffer (def. #4).

fiber cable. *See* fiber optic cable.

*****fiber cutoff wavelength (ë$_{cf}$).** *See* cutoff wavelength (def. #2).

fiber dispersion. *See* dispersion.

fiber distributed data interface (FDDI). A concept, defined in ANSI standards, for an optical-fiber-based token-ring network, featuring (a) dual counter-rotating logical rings, each with a data transmission capacity of 100 Mb/s, (b) reliable data transfer, (c) active link monitoring, (d) station management, and (e) survivability features. *Note 1:* The four standards are (a) ANSI X3T9.5, containing Physical Media Dependent (PMD) specifications, (b) ANSI X3T9.5, containing the Physical (PHY) specifications, (c) ANSI X3.139, containing Media Access Control (MAC) specifications, and (d) ANSI X39.5, containing the Station Management (SMT) specifications. *Note 2:* The data rate of an FDDI ring may be doubled to 200 Mb/s, with loss of redundancy. *Note 3:* FDDI-2, a second-generation FDDI network standard, is under development.

fiber optic bus. *See* bus.

fiber optic cable. A telecommunications cable in which one or more optical fibers are used as the propagation medium. *Note 1:* The optical fibers are surrounded by buffers, strength members, and jackets for protection, stiffness, and strength. *Note 2:* A fiber-optic cable may be an all-fiber cable, or contain both optical fibers and metallic conductors. One possible use for the metallic conductors is the transmission of electric power for repeaters. [After FAA] *Synonyms* optical cable, optical fiber cable.

fiber optic isolator. *See* optical isolator.

fiber optic link. A communications link that transmits signals by means of modulated light propagated in an optical fiber.

fiber optics (FO). The branch of optical technology concerned with the transmission of light through fibers made of transparent materials such as glasses and plastics. *Note 1:* Telecommunications applications of fiber optics use flexible low-loss fibers, using a single fiber per optical path. Present-day plastic fibers have losses that are too high for telecommunications applications. *Note 2:* Various industrial and medical applications of fiber optics, such as endoscopes, use flexible fiber bundles in which individual fibers are spatially aligned, permitting optical relay of an image. *Note 3:* Some specialized industrial applications use rigid (fused) aligned fiber bundles for image transfer; such as in the fiber optics faceplates used on some cathode ray rubes (CRTs) to "flatten" the image.

fiber pigtail. *See* pigtail (def. #1).

fidelity. The degree to which a system, or a portion of a system, accurately reproduces, at its output, the essential characteristics of the signal impressed upon its input or the result of a prescribed operation on the signal impressed upon its input.

field. 1. The volume of influence of a physical phenomenon, expressed vectorially. 2. On a data medium or in storage, a specified area used for a particular class of data, *e.g.*, a group of character positions used to enter or display wage rates on a screen. 3. Defined logical data that are part of a record. 4. The elementary unit of a record that may contain a data item, a data aggregate, a pointer, or a link. 5. In an interlaced, raster-scanned video display, a partial frame, consisting of every nth scanning line of a complete frame, where n is an integer equal to the number of fields (usually two) in a complete frame. *Note 1:* For example, in the National Television Standards Committee (NTSC) television specification used in the United States, a single frame is composed of two fields, each of which has half the number of scanning lines in a complete frame. The scanning lines of a field are separated by twice the space between the scanning lines in the full frame. The two fields are interlaced, *i.e.*, a complete frame consists of the following traces, which are listed in the order of their appearance in the complete frame, but not the order in which scanning occurs: the first line of the first field, the first line of the second field, the second line of the first field, the second line of the second field, the third line of the first field, the third line of the second field, *etc.*, until completion of the full frame. The fields are scanned alternately, one complete field at a time. Thus, the flicker rate of the display is perceived by the eye to be twice as fast as that which would result if the complete frame were to be scanned in line-by-line order. *Note 2*: Not all scanning lines are necessarily applied to user information, *i.e.*, the graphic display. Certain scanning lines, not seen under ordinary viewing conditions, are often used for transmitting test signals that indicate the quality of the displayed video.

field-disturbance sensor. A restricted radiation device which establishes a radio frequency field in its vicinity and detects changes in that field resulting from the movement of persons or objects within the radio frequency field. Examples: microwave intrusion sensors; devices that use rf energy for production line counting and sensing.

field intensity. The irradiance of an electromagnetic wave under specified conditions. *Note:* Field intensity is usually expressed in watts per square meter.

field strength. The magnitude of an electric, magnetic, or electromagnetic field at a given point. *Note:* The field strength of an electromagnetic wave is usually expressed as the rms value of the electric field, in volts per meter. The field strength of a magnetic field is usually expressed in ampere-turns per meter or in oersteds. *Synonym* radio field intensity.

field wire. A flexible insulated wire used in field telephone and telegraph systems. *Note 1:* WD-1 and WF-16 are types of field wire. *Note 2:* Field wire usually contains conductors and high-tensile-strength strands serving as strength members.

FIFO. *Abbreviation for* first-in first-out.

file. 1. The largest unit of storage structure that consists of a named collection of all occurrences in a database of records of a particular record type. 2. A set of related records treated as a unit, for example, in stock control, a file could consist of a set of invoices.

file server. 1. A high-capacity disk storage device or a computer that each computer on a network can use or access and retrieve files that can be shared among attached computers. 2. A program, running on a computer, that allows different programs, running on other computers, to access the files of that computer.

file transfer, access, and management (FTAM). An application's service and protocol based on the concept of virtual file store. *Note:* FTAM allows remote access to various levels in a file structure and provides a comprehensive set of file management capabilities.

File Transfer Protocol (FTP). *See* FTP.

fill. *See* bit stuffing.

fill bit. *See* bit stuffing.

filled cable. A cable that has a nonhygroscopic material, usually a gel, inside the jacket or sheath. *Note 1:* The nonhygroscopic material fills the spaces between the interior parts of the cable, preventing moisture from entering minor leaks in the sheath and migrating inside the cable. *Note 2:* A metallic cable, such as a coaxial cable or a metal waveguide, filled with a dielectric material, is not considered as a filled cable.

FILO. *Abbreviation for* first-in, last-out.

filter. In electronics, a device that transmits only part of the incident energy and may thereby change the spectral distribution of energy: (a) high-pass filters transmit energy above a certain frequency; (b) low-pass filters transmit energy below a certain frequency; (c) bandpass filters transmit energy of a certain bandwidth; (d) band-stop filters transmit energy outside a specific frequency band.

filtered symmetric differential phase-shift keying (FSDPSK). A method of encoding information for digital transmission in which (a) a binary 0 is encoded as a +90° change in the carrier phase and a binary 1 is encoded as a -90° change in the carrier phase, and (b) abrupt phase transitions are smoothed by filtering or other functionally equivalent pulse shaping techniques.

finished call. 1. In an information transaction, a call in which the call originator or call receiver terminates the communication and goes on hook, *i.e.,* hangs up. 2. In an information transfer transaction, the termination of the information transfer phase.

FIP. *Acronym for* Federal Information Processing.

FIP equipment. In the Federal Government, any equipment or interconnected system or subsystems of equipment (as defined in 41CFR) used in the automatic acquisition, storage, manipulation, management, movement, control, display, switching, interchange, transmission, or reception of data or information.

FIP system. In the Federal Government, any organized combination of FIP equipment, software, services, support services, or related supplies.

firmware. Software that is embedded in a hardware device that allows reading and executing the software, but does not allow modification, *e.g.,* writing or deleting data by an end user. *Note 1:* An example of firmware is a computer program in a read-only memory (ROM) integrated circuit chip. A hardware configuration is usually used to represent the software. *Note 2:* Another example of firmware is a program embedded in an erasable programmable read-only memory (EPROM) chip, which program may be modified by special external hardware, but not by an application program.

first-in first-out (FIFO). A queuing discipline in which entities in a queue leave the queue in the same order in which they arrive. *Note 1:* Service, when available, is offered to the entity that has been in the FIFO queue the longest. *Note 2:* FIFO techniques are used in message switching.

first-in last-out (FILO). A queuing discipline in which entities in a queue leave the queue in the reverse order from that in which they arrived. *Note:* An understanding of FILO techniques is important in the understanding of store-and-forward capabilities in packing switching.

first window. Of silica-based optical fibers, the transmission window at approximately 830 to 850 nm.

FISINT. *Acronym for* foreign instrumentation signals intelligence.

five-hundred (500) service. A telephone service that allows individuals to receive, via a single number, telephone calls in various locations (*e.g.*, home, office, or car phone) from call originators not necessarily using the same common carrier.

fixed access. In personal communications service (PCS), terminal access to a network in which there is a set relationship between a terminal and the access interface. *Note:* A single "identifier" serves for both the access interface and the terminal. If the terminal moves to another access interface, that terminal assumes the identity of the new interface.

fixed attenuator. *See* pad.

fixed loop. A service feature that permits an attendant on an assisted call to retain connection through the attendant position for the duration of the call. *Note:* The attendant will usually receive a disconnect signal when the call is terminated.

fixed microwave auxiliary station. A fixed station used in connection with (a) the alignment of microwave transmitting and receiving antenna systems and equipment, (b) coordination of microwave radio survey operations, and (c) cue and contact control of television pickup station operations.

fixed-reference modulation. Modulation in which the significant condition for any signal element is based on a fixed reference.

fixed-satellite service. A radiocommunication service between Earth stations at given positions when one or more satellites are used; the given position may be a specified fixed point or any fixed point within specified areas; in some cases this service includes satellite-to-satellite links, which may also be effected in the inter-satellite service, the fixed-satellite service may also include feeder links for other space radiocommunication services.

fixed service (FX). A radiocommunication service between specified fixed points.

fixed station. A station in the fixed service.

fixed storage. *Synonym* read-only storage.

fixed-tolerance-band compaction. Data compaction accomplished by storing or transmitting data only when the data fall outside prescribed limits. *Note:* An example of fixed-tolerance-band compaction in a telemetering system is the transmission of the temperature only when the temperature is above or below preestablished threshold limits. Thus, the recipient of the transmission is to assume that the value is in the prescribed range unless a signal to the contrary occurs.

flag. In data transmission or processing, an indicator, such as a signal, symbol, character, or digit, used for identification. *Note:* A flag may be a byte, word, mark, group mark, or letter that signals the occurrence of some condition or event, such as the end of a word, block, or message.

flag sequence. In data transmission or processing, a sequence of bits used to delimit, *i.e.* mark, the beginning and end of a frame. *Note 1:* An 8-bit sequence is usually used as the flag sequence; for example, the 8-bit flag sequence 01111110. *Note 2:* Flag sequences are used in bit-oriented protocols, such as Advanced Data Communication Control Procedures (ADCCP), Synchronous Data Link Control (SDLC), and High-Level Data Link Control (HDLC).

flash. A signal generated by the momentary depression of the telephone switchhook or other device. *Note:* A flash may be used to request additional services.

FLASH message. A category of precedence reserved for initial enemy contact messages or operational combat messages of extreme urgency. Brevity is mandatory.

flat fading. Fading in which all frequency components of a received radio signal vary in the same proportion simultaneously.

flat rate service. Telephone service in which a single payment permits an unlimited number of local calls to be made without further charge for a specified period of time.

flat weighting. In a noise-measuring set, a noise weighting based on an amplitude-frequency characteristic that is flat over a frequency range that must be stated. *Note 1:* Flat noise power is expressed in dBrn (f_1 - f_2) or in dBm (f_1 - f_2). *Note 2: "3-kHz flat weighting"* and *"15-kHz flat weighting"* are based on amplitude-frequency characteristics that are flat between 30 Hz and the frequency indicated.

F layer. *See* F region.

Fleming's rule. A rule stating that if the thumb of the right hand points in the direction of an electric current, then the curled fingers point in the direction of the magnetic field that encircles the current; and further, if the curled fingers of the right hand describe the electric current in a solenoid, then the thumb points in the direction of the magnetic field inside the solenoid. *Synonym* right-hand rule.

flexible disk. *Synonym* diskette.

flip-flop. A device that may assume either one of two reversible, stable states. *Note 1:* The flip-flop is used as a basic control element in computer and communications systems. *Note 2:* In a flip-flop, the transition from one stable state to the other is unstable, *i.e.,* for the very short period during which the transition takes place, both outputs may assume the same state, which state may be unpredictable. *Synonyms* bistable circuit, bistable multivibrator, bistable trigger circuit.

floating-point coding compaction. Data compaction accomplished by using coefficients, a base, and exponents to specify the scale, range, or magnitude of numbers. *Note:* An example of floating-point coding compaction is using 119.8×10^6, 119.8(6), or 119.86 to represent 119,800,000. If the number is rounded to 120,000,000, it might be written as 1206 or 127 in which the last digit is the number of zeros to be appended to the preceding digits. Thus, only three positions are required instead of nine to represent the number in storage or in a message, which is only 33% of the original space and time requirement.

flooding compound. A substance surrounding the buffer tubes of a fiber-optic cable, to prevent water intrusion into the interstices in the event of a breach of the jacket.

flood projection. In facsimile, the optical method of scanning in which the object is floodlighted and the scanning spot is defined by a masked portion of the illuminated area.

flood search routing. In a telephone network, nondeterministic routing in which a dialed number received at a switch is transmitted to all switches, *i.e.,* flooded, in the area code directly connected to that switch; if the dialed number is not an affiliated subscriber at that switch, the number is then retransmitted to all directly connected switches, and then routed through the switch that has the dialed number corresponding to the particular user end instrument affiliated with it. *Note 1:* All digits of the numbering plan are used to identify a particular subscriber. *Note 2:* Flood search routing allows subscribers to have telephone numbers independent of switch codes. *Note 3:* Flood search routing provides the highest probability that a call will go through even though a number of switches and links fail.

floppy disk. *Synonym* diskette.

flops. *Acronym for* floating-point operations per second. *Note:* For example, 15 Mflops equals 15 million floating-point arithmetic operations per second.

flowchart. A graphical representation in which symbols

are used to represent such things as operations, data, flow direction, and equipment, for the definition, analysis, or solution of a problem. *Synonym* flow diagram.

flow control. *See* transmit flow control.

flow control procedure. A procedure for controlling the rate of transfer of data among elements of a network, *e.g.*, between a DTE and a data switching exchange network, to prevent overload.

flow diagram. *Synonym* flowchart.

flowline. On a flowchart, a line that (a) has an indicated direction, (b) represents a connection between other symbols, and (c) indicates the sequence of operations or the transfer of control.

flutter. Rapid variation of signal parameters, such as amplitude, phase, and frequency. *Note:* Examples of flutter are (a) rapid variations in received signal levels, such as variations that may be caused by atmospheric disturbances, antenna movements in a high wind, or interaction with other signals, (b) in radio propagation, a phenomenon in which nearly all radio signals that are usually reflected by ionospheric layers in or above the E-region experience partial or complete absorption, (c) in radio transmission, rapidly changing signal levels, together with variable multipath time delays, caused by reflection and possible partial absorption of the signal by aircraft flying through the radio beam or common scatter volume, (d) the variation in the transmission characteristics of a loaded telephone circuit caused by the action of telegraph direct currents on the loading coils, (e) in recording and reproducing equipment, the deviation of frequency caused by irregular mechanical motion, *e.g.,* that of capstan angular velocity in a tape transport mechanism, during operation.

flux. 1. The lines of force of a magnetic field. 2. *Obsolete synonym for* radiant power.

flywheel effect. In an oscillator, the continuation of oscillations after removal of the control stimulus. *Note 1:* The flywheel effect is usually caused by interacting inductive and capacitive circuits in the oscillator. *Note 2:* The flywheel effect may be desirable, such as in phase-locked loops used in synchronous systems, or undesirable, such as in voltage-controlled oscillators. *Synonym* flywheeling.

flywheeling. *Synonym* flywheel effect.

FM. *Abbreviation for* frequency modulation.

FM blanketing. That form of interference to the reception of other broadcast stations, which is caused by the presence of an FM broadcast signal of 115 dBμ (562 mV/m) or greater signal strength in the area adjacent to the antenna of the transmitting station. The 115-dBu contour is referred to as the *"blanketing area."*

FM broadcast translator. *See* translator (def. #3).

FM capture effect. *Synonym* capture effect.

FM capture ratio. *See* capture effect.

FM improvement factor. The quotient obtained by dividing the signal-to-noise ratio (SNR) at the output of an FM receiver by the carrier-to-noise ratio (CNR) at the input of the receiver. *Note:* When the FM improvement factor is greater than unity, the improvement in the SNR is always obtained at the expense of an increased bandwidth in the receiver and the transmission path.

FM improvement threshold. The point in an FM receiver at which the peaks in the rf signal equal the peaks of the thermal noise generated in the receiver. *Note:* A baseband signal-to-noise ratio of about 30 dB is typical at the improvement threshold, and this ratio improves 1 dB for each decibel of increase in the signal above the threshold.

FM threshold effect. In an FM receiver, the effect produced when the desired-signal gain begins to limit the desired signal, and thus noise limiting (suppression). *Note:* FM threshold effect occurs at (and above) the point at which the FM signal-to-noise improvement is measured.

FM threshold extension. A change in the value of the FM threshold of a receiver. *Note:* FM threshold extension may be obtained by decreasing the operational bandwidth, thus decreasing the received noise power and allowing the threshold of the desired signal to occur at a lower signal input level.

FO. *Abbreviation for* fiber optics.

footprint. In satellite communications, that portion of the Earth's surface over which a satellite antenna delivers a specified amount of signal power under specified conditions. *Note:* The limiting case of footprint area is somewhat less than one-half the Earth's surface, and depends on the altitude of the satellite.

forbidden character. *Synonym* illegal character.

foreign exchange (FX) service. A network-provided service in which a telephone in a given local exchange area is connected, via a private line, to a central office in another, *i.e., "foreign"*, exchange, rather than the local exchange area's central office. *Note:* To call originators, it appears that the subscriber having the FX service is located in the foreign exchange area.

foreign instrumentation signals intelligence (FISINT). 1. Intelligence information derived from electromagnetic emissions associated with the testing and operational deployment of foreign aerospace, surface, and subsurface systems. 2. Technical information and intelligence information derived from the intercept of foreign instrumentation signals by other than the intended recipients. Foreign instrumentation signals intelligence is a category of signals intelligence. *Note:* Foreign instrumentation signals include but are not limited to signals from telemetry, beaconry, electronic interrogators, track-ing/fusing/arming/firing command systems, and video data links.

format. 1. The arrangement of bits or characters within a group, such as a word, message, or language. 2. The shape, size, and general makeup of a document.

Fortran. *See* language processor.

fortuitous conductor. Any conductor that may provide an unintended path for signals. *Note:* Examples of fortuitous conductors are water pipes, wires, cables, and metal building and equipment structural members.

fortuitous distortion. Distortion resulting from causes generally subject to laws concerning random occurrences.

forward busying. In a telecommunications system, a feature in which supervisory signals are forwarded in advance of address signals in order to seize assets of the system before attempting to establish a connection.

forward channel. The channel of a data circuit that transmits data from the originating user to the destination user. *Note:* The forward channel carries message traffic and some control information.

forward echo. In a transmission line, an echo propagating in the same direction as the original wave and consisting of energy reflected back by one discontinuity and then forward again by another discontinuity. *Note:*

Forward echoes can be supported by reflections caused by splices or other discontinuities in the transmission medium (*e.g.*, optical fiber, twisted pair, or coaxial tube). In metallic lines, they may be supported by impedance mismatches between the source or load and the characteristic impedance of the transmission medium.

forward error correction (FEC). A system of error control for data transmission wherein the receiving device has the capability to detect and correct any character or code block that contains fewer than a predetermined number of symbols in error. *Note:* FEC is accomplished by adding bits to each transmitted character or code block, using a predetermined algorithm.

forward propagation ionospheric scatter (FPIS). *Synonym* ionospheric scatter.

forward scatter. The deflection—by diffraction, nonhomogeneous refraction, or nonspecular reflection by particulate matter of dimensions that are large with respect to the wavelength in question but small with respect to the beam diameter—of a portion of an incident electromagnetic wave, in such a manner that the energy so deflected propagates in a direction that is within 90° of the direction of propagation of the incident wave. *Note:* The scattering process may be polarization-sensitive, *i.e.*, incident waves that are identical in every respect but their polarization may be scattered differently.

forward signal. A signal sent in the direction from the calling to the called station, *i.e.*, from the original data source to the original data sink. *Note:* The forward signal is transmitted in the forward channel.

FOT. *Abbreviation for* frequency of optimum transmission. In the transmission of radio waves via ionospheric reflection, the highest effective, *i.e.*, working, frequency that is predicted to be usable for a specified path and time for 90% of the days of the month. *Note 1:* The FOT is normally just below the value of the maximum usable frequency (MUF). In the prediction of usable frequencies, the FOT is commonly taken as 15% below the monthly median value of the MUF for the specified time and path. *Note 2:* The FOT is usually the most effective frequency for ionospheric reflection of radio waves between two specified points on Earth. *Synonyms* frequency of optimum traffic, optimum traffic frequency, optimum transmission frequency, optimum working frequency.

Fourier analysis. The definition of a periodic waveform of arbitrary shape as a summation of sine waves having specific amplitudes and phases, and having frequencies corresponding to the harmonics of the waveform being defined. *Note:* A Fourier analysis is particularly well suited for communications equipment design and for predicting the performance of a given design.

four-wire circuit. A two-way circuit using two paths so arranged that the respective signals are transmitted in one direction only by one path and in the other direction by the other path. *Note:* The four-wire circuit gets its name from the fact that, historically, two conductors were used in each of two directions for full-duplex operation. The name may still be applied, *e.g.*, to a communications link supported by optical fibers, even though only one fiber is required for transmission in each direction. *Contrast with* two-wire circuit.

four-wire repeater. A repeater, consisting of two amplifiers, one associated with each direction, used in a four-wire circuit.

four-wire terminating set. A balanced transformer used to perform a conversion between 4-wire and 2-wire operation. *Note 1:* For example, a 4-wire circuit may, by means of a 4-wire terminating set, be connected to a 2-wire telephone set. Also, a pair of 4-wire terminating sets may be used to introduce an intermediate 4-wire loop into a 2-wire circuit, in which loop repeaters may be situated to amplify signals in each direction without positive feedback and oscillation. *Note 2*: Four-wire terminating sets have been largely supplanted by resistance hybrids.

fox message. A standard test message that includes all the alphanumerics on a teletypewriter and also function characteristics (space, figures shift, letters shift). *Note:* An example of a fox message is "THE QUICK BROWN FOX JUMPED OVER THE LAZY DOG'S BACK 1234567890."

FPIS. *Abbreviation for* forward propagation ionospheric scatter. *See* ionospheric scatter.

fractional frequency fluctuation. The deviation of the frequency of an oscillator from its nominal constant frequency, normalized to the nominal frequency.

fractional T1. In the North American or Japanese hierarchies, the tariffed use of a data rate corresponding to fewer than the 24 channels served by a T1 line.

frame. 1. In data transmission, the sequence of contiguous bits delimited by, and including, beginning and ending flag sequences. *Note 1:* A frame usually includes an information field, and usually consists of a specified number of bits between flags and contains an address field, a control field, a frame check sequence, and flags. *Note 2:* Frames usually consist of a representation of the original data to be transmitted, together with other bits which may be used for error detection or control.

Additional bits may be used for routing, synchronization, or overhead information not directly associated with the original data. 2. In the multiplex structure of pulse-code modulation (PCM) systems, a set of consecutive time slots in which the position of each digit can be identified by reference to a frame-alignment signal. *Note:* The frame-alignment signal does not necessarily occur, in whole or in part, in each frame. 3. In a time-division multiplexing (TDM) system, a repetitive group of signals resulting from a single sampling of all channels, including any required system information, such as additional synchronizing signals. *Note: "In-frame"* is the condition that exists when there is a channel-to-channel and bit-to-bit correspondence, exclusive of transmission errors, between all inputs of a time-division multiplexer and the output of its associated demultiplexer. 4. In ISDN, a block of variable length, labeled at the Data Link Layer of the Open Systems Interconnection—Reference Model. 5. In video display, the set of all picture elements that represent one complete image. *Note:* In NTSC and other television standards used throughout the world, a frame consists of two interlaced fields, each of which has half the number of scanning lines, and consequently, half the number of pixels, of one frame. 6. In video display, one complete scanned image from a series of video images. *Note:* A video frame is usually composed of two interlaced fields.

frame alignment. In the reception of framed digital data, the extent to which a received frame is correctly aligned with respect to the clock at the receiver.

frame-alignment recovery time. *Synonym* reframing time.

frame alignment signal. In the transmission of data frames, a distinctive sequence of bits used to accomplish frame alignment. *Note:* A frame alignment signal may also contain additional bits for status, control, and error detection.

frame-alignment time slot . A time slot starting at a particular phase or instant in each frame and allocated to the transmission of a frame-alignment signal.

frame check sequence (FCS). *See* cyclic redundancy check.

framed interface. An interface through which information flow is partitioned into physical, periodic frames consisting of overhead information and an information payload.

frame duration. The time between the beginning of a frame and the end of that frame. *Note:* For fixed-length frames, at a fixed data rate, frame duration is constant.

frame frequency. *Synonym* frame rate.

frame grabber. A device that can seize and record a single frame of video information out of a sequence of many frames.

frame pitch. The distance, time, or number of bits between corresponding points, *i.e.,* significant instants, in two consecutive frames.

frame rate. The number of frames transmitted or received per unit time. *Note 1:* The frame rate is usually expressed in frames per second. *Note 2:* In television transmission, the frame rate must be distinguished from the field rate, which in the NTSC and other systems, is twice the frame rate. *Synonym* frame frequency.

frame relay. An interface protocol for statistically multiplexed packet-switched data communications in which (a) variable-sized packets (frames) are used that completely enclose the user packets they transport, and (b) transmission rates are usually between 56 kb/s and 1.544 Mb/s (the T-1 rate). *Note 1:* In frame relay, (a) there is neither flow-control nor an error-correction capability, (b) there is information-content independence, (c) there is a correspondence only to the ISO Open systems Interconnection—Reference Model Layers 1 and 2, (d) variable-sized user packets are enclosed in larger packets (frames) that add addressing and verification information, (e) frames may vary in length up to a design limit, usually 1 kilobyte or more, (f) one frame relay packet transports one user packet, (g) implementation of fast-packet technology is used for connection-oriented frame relay services, and (h) there is a capability to handle time-delay insensitive traffic, such as LAN interworking and image transfer. *Note 2:* Frame relay is referred to as the *local management interface (LMI) standard* and is specified in *ANSI T1.617.*

frame slip. In the reception of framed data, the loss of synchronization between a received frame and the receiver clock, causing a frame misalignment event, and resulting in the loss of the data contained in the received frame. *Note:* A frame slip should not be confused with a dropped frame where synchronization is not lost, *e.g.*, in the case of buffer overflow.

frame synchronization. Of a received stream of framed data, the process by which incoming frame alignment signals, *i.e.*, distinctive bit sequences, are identified, *i.e.*, distinguished from data bits, permitting the data bits within the frame to be extracted for decoding or retransmission. *Note:* The usual practice is to insert, in a dedicated time slot within the frame, a noninformation bit that is used for the actual synchronization of the incoming data with the receiver. *Synonym* framing (def. #1).

frame synchronization pattern. In digital communications, a prescribed recurring pattern of bits transmitted to enable the receiver to achieve frame synchronization.

framing. 1. In time-division multiplexing reception, *synonym* frame synchronization. 2. In video reception, the process of adjusting the timing of the receiver to coincide with the received video synchronization pulse. 3. In facsimile, the adjustment of the facsimile picture to a desired position in the direction of line progression.

framing bit. 1. A bit used for frame synchronization. 2. In a bit stream, a bit used in determining the beginning or end of a frame. *Note 1:* The framing bit occurs at a specific position in the frame. *Note 2:* In a bit stream, framing bits are noninformation bits. *Note 3:* Framing in a digital signal is usually repetitive.

framing signal. *See* frame-alignment signal, framing bit.

Fraunhofer diffraction pattern. *Synonym* far-field diffraction pattern.

Fraunhofer region. *Synonym* far-field region.

free net. A radio net in which any station may communicate with any other station in the net without first obtaining the permission of the net-control station. *Note:* Permission to operate as a free net is granted by the net-control station until such time as a directed net is established by the net-control station.

free routing. The routing of messages in such a manner that they are forwarded toward their destination or addressee over any available channel without dependence upon predetermined routing.

free-running capability. In a synchronized oscillator, the capability to operate in the absence of a synchronizing signal.

free space. A theoretical concept of space devoid of all matter. *Note:* Free space also implies remoteness from material objects that could influence the propagation of electromagnetic waves.

free-space coupling. Coupling of magnetic, electric, or electromagnetic fields that are not confined to a conductor. *Note:* Coupling by the deliberate introduction of capacitors and inductors is not considered free-space coupling.

free-space loss. The signal attenuation that would result if all absorbing, diffracting, obstructing, refracting, scattering, and reflecting influences were sufficiently removed so as to have no effect on propagation. *Note:* Free-space loss is primarily caused by beam divergence, *i.e.*, signal energy spreading over larger areas at increased distances from the source.

freeze frame. A frame of visual information that is selected from a set of motion video frames, and is held in a buffer. *Contrast with* still video.

freeze frame television. Television in which fixed ("still") images are transmitted sequentially at a rate far too slow to be perceived as continuous motion by human vision. *Note:* Transmission of an image is usually performed periodically by a processing unit that contains memory in which data representing the image are stored. For an image of specified quality, *e.g.*, resolution and color fidelity, freeze-frame television has a lower bandwidth requirement than that of full-motion television.

F region. That portion of the ionosphere existing between approximately 160 and 400 km above the surface of the Earth, consisting of layers of increased free-electron density caused by the ionizing effect of solar radiation. *Note 1:* The F region reflects normal-incident frequencies at or below the critical frequency (approximately 10 MHz) and partially absorbs waves of higher frequency. *Note 2:* The F_1 layer exists from about 160 to 250 km above the surface of the Earth and only during daylight hours. Though fairly regular in its characteristics, it is not observable everywhere or on all days. The principal reflecting layer during the summer for paths of 2,000 to 3,500 km is the F_1 layer. The F_1 layer has approximately 5×10^5 e/cm^3 (free electrons per cubic centimeter) at noontime and minimum sunspot activity, and increases to roughly 2×10^6 e/cm^3 during maximum sunspot activity. The density falls off to below 10^4 e/cm^3 at night. *Note 3:* The F_1 layer merges into the F_2 layer at night. *Note 4:* The F_2 layer exists from about 250 to 400 km above the surface of the Earth. The F_2 layer is the principal reflecting layer for HF communications during both day and night. The horizon-limited distance for one-hop F_2 propagation is usually around 4,000 km. The F_2 layer has about 10^6 e/cm^3. However, variations are usually large, irregular, and particularly pronounced during magnetic storms.

frequency. For a periodic function, the number of cycles or events per unit time.

frequency accuracy. The degree of conformity to a specified value of a frequency.

frequency aging. Of an oscillator, the change in frequency, over time, caused by internal changes in oscillator parameters even when external factors, such as environment and power supply characteristics, are constant.

frequency allocation. *See* allocation (of a frequency band).

frequency allotment. *See* allotment (of a radio frequency or radio frequency channel).

frequency-analysis compaction. Data compaction accomplished by using an expression composed of a number of different frequencies of different magnitudes to represent a particular curve. *Note:* An example of frequency-analysis compaction is the use of a Fourier analysis to represent an arbitrary curve, a periodic function, an aperiodic function, or a wave shape. Thus, the fundamental frequency, the amplitude of the fundamental frequency, and the amplitudes and frequencies of the harmonics are all that are needed to reconstitute the function or wave shape. The shape can thus be readily stored and transmitted in this compacted form. [From Weik '89]

frequency assignment. 1. Authorization, given by an Administration, for a radio station to use a radio frequency or radio frequency channel under specified conditions 2. The process of authorizing a specific frequency, group of frequencies, or frequency band to be used at a certain location under specified conditions, such as bandwidth, power, azimuth, duty cycle, or modulation. *Synonym* radio frequency channel assignment. *See* administration (def. #1).

frequency assignment authority. The power granted an Administration, or its designated or delegated leader or agency via treaty or law, to specify frequencies, or frequency bands, in the electromagnetic spectrum for use in systems or equipment. *Note:* Primary frequency assignment authority for the United States is exercised by the National Telecommunication's and Information Administration (NTIA) for the Federal Government and by the Federal Communications Commission (FCC) for non-Federal Government organizations. International frequency assignment authority is vested in the International Frequency Registration Board of the International Telecommunication Union.

frequency averaging. 1. The process by which the relative phases of precision clocks are compared for the purpose of defining a single time standard. 2. A process in which network synchronization is achieved by use, at all nodes, of oscillators that adjust their frequencies to the average frequency of the digital bit streams received from connected nodes. *Note:* In frequency averaging, all oscillators are assigned equal weight in determining the ultimate network frequency.

frequency band. *See* electromagnetic spectrum.

frequency band allocation. *See* allocation (of a frequency band).

frequency-change signaling. A signaling method in which one or more discrete frequencies correspond to each desired significant condition of a code. *Note 1:* The transition from one set of frequencies to the other may be a continuous or a discontinuous change in frequency or in phase. *Note 2:* Frequency-change signaling may be used in both supervisory signaling and data transmission.

frequency coherence. *See* phase coherence.

frequency compatibility. 1. Of an electronic device, the extent to which it will operate at its designed performance level in its intended operational environment (including the presence of interference) without causing interference to other devices. 2. The degree to which an electrical or electronic device or devices operating on or responding to a specified frequency or frequencies is capable of functioning with other such devices.

frequency departure. An unintentional deviation from the nominal frequency value.

frequency-derived channel. A channel derived by dividing an allocated or available bandwidth over a medium into two or more portions, each usable separately. *Note:* A frequency-derived channel is continuously available and may be further divided on either a frequency or time basis.

frequency deviation. 1. The amount by which a frequency differs from a prescribed value, such as the amount an oscillator frequency drifts from its nominal frequency. 2. In frequency modulation, the absolute difference between (a) the maximum permissible instantaneous frequency of the modulated wave or the minimum permissible instantaneous frequency of the modulated wave and (b) the carrier frequency. 3. In frequency modulation, the maximum absolute difference, during a specified period, between the instantaneous frequency of the modulated wave and the carrier frequency.

frequency dispersal. An electronic counter-countermeasure (ECCM) in which communications nets' operating frequencies are widely separated from each other, causing a requirement to spread jamming power over wider frequency bands and thus compelling a reduction of available jamming power on any single channel or frequency, or causing a requirement for more jamming power or more jamming equipment.

frequency displacement. The end-to-end shift in frequency that may result from independent frequency translation errors in a circuit.

frequency distortion. *Synonym* amplitude-vs.-frequency distortion.

frequency diversity. Transmission and reception in which the same information signal is transmitted and received simultaneously on two or more independently fading carrier frequencies.

frequency-division multiple access (FDMA). The use of frequency division to provide multiple and simultaneous transmissions to a single transponder.

frequency-division multiplexing (FDM). The deriving of two or more simultaneous, continuous channels from a transmission medium by assigning a separate portion of the available frequency spectrum to each of the individual channels.

frequency drift. An undesired progressive change in frequency with time. *Note 1:* Causes of frequency drift include component aging and environmental changes. *Note 2:* Frequency drift may be in either direction and is not necessarily linear.

frequency-exchange signaling. Frequency-change signaling in which the change from one significant condition to another is accompanied by decay in amplitude of one or more frequencies and by buildup in amplitude of one or more other frequencies. *Note:* Frequency-exchange signaling applies to supervisory signaling and user-information transmission. *Synonym* two-source frequency keying.

frequency fluctuation. A short-term variation, with respect to time, of the frequency of an oscillator. *Note:* Frequency fluctuation, $f(t)$, is given by

$$f(t) = \frac{1}{2\pi}\frac{d^2\,\theta(t)}{dt^2},$$

where $\theta(t)$ is the phase angle of the sinusoidal wave with respect to time, t.

frequency frogging. 1. The interchanging of the frequencies of carrier channels to accomplish specific purposes, such as to prevent feedback and oscillation, to reduce crosstalk, and to correct for a high frequency-response slope in the transmission line. *Note:* Frequency frogging is accomplished by having modulators, which are integrated into specially designed repeaters, translate a low-frequency group to a high-frequency group, and vice versa. A channel will appear in the low group for one repeater section and will then be translated to the high group for the next section because of frequency frogging. This results in nearly constant attenuation with frequency over two successive repeater sections, and eliminates the need for large slope equalization and adjustments. Singing and crosstalk are minimized because the high-level output of a repeater is at a different frequency than the low-level input to other repeaters. 2. In microwave systems, the alternate use of two frequencies at repeater sites to prevent feedback and oscillation.

frequency guard band. A frequency band deliberately left vacant between two channels to provide a margin of safety against mutual interference.

frequency hopping. [The] repeated switching of frequencies during radio transmission according to a specified algorithm, to minimize unauthorized interception or jamming of telecommunications. *Note:* The overall bandwidth required for frequency hopping is much wider than that required to transmit the same information using only one carrier frequency.

frequency-hopping spread spectrum. A signal structuring technique employing automatic switching of the transmitted frequency. Selection of the frequency to be transmitted is typically made in a pseudo-random manner from a set of frequencies covering a band wider than the information bandwidth. The intended receiver would frequency-hop in synchronization with the code of the transmitter in order to retrieve the desired information. *Note:* In many cases, used as an electronic counter-countermeasure (ECCM) technique.

frequency hour. One frequency used for one hour regardless of the number of transmitters over which it is simultaneously broadcast by a station during that hour.

frequency instability. *See* **frequency stability.**

frequency lock. The condition in which a frequency-correcting feedback loop maintains control of an oscillator within the limits of one cycle. *Note:* Frequency lock does not imply phase lock, but phase lock does imply frequency lock.

frequency modulation (FM). Modulation in which the instantaneous frequency of a sine wave carrier is caused to depart from the center frequency by an amount proportional to the instantaneous value of the modulating signal. *Note 1:* In FM, the carrier frequency is called the center frequency. *Note 2:* FM is a form of angle modulation. *Note 3:* In optical communications, even if the electrical baseband signal is used to frequency-modulate an electrical carrier (an "FM" optical communications system), it is still the intensity of the lightwave that is varied (modulated) by the electrical FM carrier. In this case, the "information,"as far as the lightwave is concerned, is the electrical FM carrier. The lightwave is varied in intensity at an instantaneous rate corresponding to the instantaneous frequency of the electrical FM carrier.

frequency offset. The difference between the frequency of a source and a reference frequency.

frequency of optimum traffic (FOT). *Synonym* FOT.

frequency of optimum transmission (FOT). *See* FOT.

frequency prediction. A prediction of the maximum usable frequency (MUF), the optimum traffic frequency, and the lowest usable frequency (LUF) for transmission between two specific locations or geographical areas during various times throughout a 24-hour period. *Note:* The prediction is usually indicated by means of a graph for each frequency plotted as a function of time.

frequency range. A continuous range or spectrum of frequencies that extends from one limiting frequency to another. *Note 1:* The frequency range for given equipment specifies the frequencies at which the equipment is operable. For example, filters pass or stop certain bands of frequencies. The frequency range for propagation indicates the frequencies at which electromagnetic wave propagation in certain modes or paths is possible over given distances. Frequency allocation, however, is made in terms of bands of frequencies. There is little, if any, conceptual difference between a range of frequencies and a band of frequencies. *Note 2: "Frequency band"* usually identifies a specific band of frequencies in the Tables of Frequency Allocations.

frequency response. *See* insertion-loss-vs.-frequency characteristic.

frequency response curve. A plot of the gain or attenuation of a device, such as an amplifier or a filter, as a function of frequency. *Note:* A flat curve indicates a uniform gain or attenuation over the range of frequencies for which the curve is flat. Most amplifiers have a flat frequency response over a certain band, above and below which the gain is reduced. The frequency response curve of a filter has one or more peaks or troughs.

frequency scanning. Conducting a search for signals over a band or range of frequencies by means of a manually or automatically tuned receiver. *Note:* The tuning rate, *i.e.,* the frequency change rate, may be fixed or variable, or it may be performed mechanically at low speed or electronically at high speed. Frequency scanning may be used to enable a radar to transmit on a clear frequency, *i.e.*, a no-interference frequency, by searching a frequency band and then tuning the system to a clear portion of that band.

frequency sharing. The assignment to or use of the same radio frequency by two or more stations that are separated geographically or that use the frequency at different times. *Note 1:* Frequency sharing reduces the potential for mutual interference where the assignment of different frequencies to each user is not practical or possible. *Note 2:* In a communications net, frequency sharing does not pertain to stations that use the same frequency.

frequency shift. 1. Any change in frequency. 2. Any change in the frequency of a radio transmitter or oscillator. *Note:* In the radio regime, frequency shift is also called rf shift. 3. *See* frequency-shift telegraphy. 4. In facsimile, a frequency modulation system where one frequency represents picture black and another frequency represents picture white. Frequencies between these two limits may represent shades of gray. 5. An intentional frequency change used for modulation purposes.

frequency-shift keying (FSK). Frequency modulation in which the modulating signal shifts the output frequency between predetermined values. *Note 1:* Usually, the instantaneous frequency is shifted between two discrete values termed the *"mark"* and *"space"* frequencies. This is a noncoherent form of FSK. *Note 2:* Coherent forms of FSK exist in which there is no phase discontinuity in the output signal. *Synonyms* frequency-shift modulation, frequency-shift signaling.

frequency-shift modulation. *Synonym* frequency-shift keying.

frequency-shift signaling. *Synonym* frequency-shift keying.

frequency-shift telegraphy. Telegraphy by frequency modulation in which the telegraph signal shifts the frequency of the carrier between predetermined values.

frequency source. *See* frequency standard.

frequency spectrum. *See* electromagnetic spectrum.

frequency spectrum congestion. The situation that occurs when many stations transmit simultaneously using frequencies that are close together, *i.e.,* with insufficient width of frequency guard bands or channel spacing. *Note:* Frequency spectrum congestion causes (a) difficulty in discrimination by tuning, (b) overlap of (i) a sideband and an adjacent carrier, or (ii) upper and lower sidebands, respectively, of adjacent carriers, and (c) interference that occurs when frequencies shift slightly or are phase shifted by ionospheric reflection.

frequency stability. The degree to which variations of the frequency of an oscillator deviate from the mean frequency over a specified period of time.

frequency standard. A stable oscillator used for frequency calibration or reference. *Note 1:* A frequency standard generates a fundamental frequency with a high degree of accuracy and precision. Harmonics of this fundamental frequency are used to provide reference points. *Note 2:* Frequency standards in a network or facility are sometimes administratively designated as "primary" or "secondary." The terms *"primary"* and *"secondary,"* as used in this context, should not be confused with the respective technical meanings of these words in the discipline of precise time and frequency.

frequency synthesizer. A device that produces frequencies that are phase coherent with a reference frequency. *Note:* The reference frequency may be derived from an internal or external source.

frequency tolerance. The maximum permissible departure by the center frequency of the frequency band occupied by an emission from the assigned frequency, or by the characteristic frequency of an emission from the reference frequency. Frequency tolerance is expressed in parts per 10^6 or in hertz. *Note:* In the United States, frequency tolerance is expressed in parts per 10^n, in hertz, or in percentages. Frequency tolerance includes both the initial setting tolerance and excursions related to short- and long-term instability and aging.

frequency translation. The transfer of signals occupying a specified frequency band, such as a channel or group of channels, from one portion of the frequency spectrum to another, in such a way that the arithmetic frequency difference of signals within the band is unaltered.

Fresnel diffraction pattern. *Synonym* near-field diffraction pattern.

Fresnel reflection. In optics, the reflection of a portion of incident light at a discrete interface between two media having different refractive indices. *Note 1:* Fresnel reflection occurs at the air-glass interfaces at the entrance and exit ends of an optical fiber. Resultant transmission losses, on the order of 4% per interface, can be reduced considerably by the use of index-matching materials. *Note 2:* The coefficient of reflection depends upon the refractive index difference, the angle of incidence, and the polarization of the incident radiation. For a normal ray, the fraction of reflected incident power is given by

$$R = \frac{(n_1 - n_2)^2}{(n_1 + n_2)^2},$$

where R is the reflection coefficient and n_1 and n_2 are the respective refractive indices of the two media. In general, the greater the angle of incidence with respect to the normal, the greater the Fresnel reflection coefficient, but for radiation that is linearly polarized in the plane of incidence, there is zero reflection at Brewster=s angle. *Note 3:* Macroscopic optical elements may be given antireflection coatings consisting of one or more dielectric thin-film layers having specific refractive indices and thicknesses. Antireflection coatings reduce overall Fresnel reflection by mutual interference of individual Fresnel reflections at the boundaries of the individual layers.

Fresnel zone. In radio communications, one of a (theoretically infinite) number of a concentric ellipsoids of revolution which define volumes in the radiation pattern of a (usually) circular aperture. *Note 1:* The cross section of the first Fresnel zone is circular. Subsequent Fresnel zones are annular in cross section, and concentric with the first. *Note 2:* Odd-numbered Fresnel zones have relatively intense field strengths, whereas even numbered Fresnel zones are nulls. *Note 3:* Fresnel zones result from diffraction by the circular aperture.

front-end noise temperature. A measure of the thermal noise generated in the first stage of a receiver.

front-end processor (FEP). A programmed-logic or stored-program device that interfaces data communication equipment with an input/output bus or memory of a data processing computer.

front-to-back ratio. 1. Of an antenna, the gain in a specified direction, *i.e.*, azimuth, usually that of maximum gain, compared to the gain in a direction 180° from the specified azimuth. *Note:* Front-to-back ratio is usually expressed in dB. 2. A ratio of parameters used to characterize rectifiers or other devices, in which electrical current, signal strength, resistance, or other parameters, in one direction is compared with that in the opposite direction.

FSDPSK. *Abbreviation for* filtered symmetric differential phase-shift keying.

FSK. *Abbreviation for* frequency-shift keying.

FTAM. *Abbreviation for* file transfer, access, and management.

FTP. *Abbreviation for* File Transfer Protocol. The Transmission Control Protocol/Internet Protocol (TCP/IP) protocol that is (a) a standard high-level protocol for transferring files from one computer to another, (b) usually implemented as an application level program, and (c) uses the Telnet and TCP protocols. *Note:* In conjunction with the proper local software, FTP allows computers connected to the Internet to exchange files, regardless of the computer platform.

FTS. *See* FTS2000.

FTS2000. *Abbreviation for* Federal Telecommunications System 2000. A long distance telecommunications service, including services such as switched voice service for voice or data up to 4.8 kb/s, switched data at 56 kb/s and 64 kb/s, switched digital integrated service for voice, data, image, and video up to 1.544 Mb/s, packet switched service for data in packet form, video transmission for both compressed and wideband video, and dedicated point-to-point private line for voice and data. *Note 1:* Use of FTS2000 contract services is mandatory for use by U.S. Government agencies for all acquisitions subject to 40 U.S.C. 759. *Note 2:* No U.S. Government information processing equipment or customer premises equipment other than that which are required to provide an FTS2000 service are furnished. *Note 3:* The FTS2000 contractors will be required to provide service directly to an agency's terminal equipment interface. For example, the FTS2000 contractor might provide a terminal adapter to an agency location in order to connect FTS2000 ISDN services to the agency's terminal equipment. *Note 4:* GSA awarded two 10-year, fixed-price contracts covering FTS2000 services on December 7, 1988. *Note 5:* The Warner Amendment excludes the mandatory use of FTS2000 in instances related to maximum security.

full carrier. A carrier that is transmitted without reduction in power, *i.e.*, a carrier that is of sufficient level to demodulate the sideband(s).

full carrier single-sideband emission. A single-sideband emission without reduction of the carrier.

full-duplex (FDX) circuit. A circuit that permits simultaneous transmission in both directions.

full-duplex (FDX) operation. *Synonym* duplex operation.

full duration at half maximum (FDHM). Full width at half maximum in which the independent variable is time. *See* full width at half maximum.

full modulation. In an analog-to-digital converter, the condition in which the input signal amplitude has just reached the threshold at which clipping begins to occur.

full-motion operation. In television, a video frame update rate that provides the appearance of full motion without flicker or smear problems. *Note:* Picture motion appears to be full at greater than 16 fps (frames per second). European television operates at 25 fps and North American television at 30 fps.

full processing. All processing functions required to recover the information bits from a received signal.

full width at half maximum (FWHM). An expression of the extent of a function, given by the difference between the two extreme values of the independent variable at which the dependent variable is equal to half of its maximum value. *Note 1:* FWHM is applied to such phenomena as the duration of pulse waveforms and the spectral width of sources used for optical communications. *Note 2:* The term *full duration at half maximum (FDHM)* is preferred when the independent variable is time.

fully connected mesh network. *See* network topology.

fully connected topology. *See* network topology.

fully intermateable connectors. Connectors from one source that mate with complementary components from other sources without mechanical damage and with transmission properties maintained within specified limits.

functional component (FC). In intelligent networks, an elemental call-processing component that directs internal network resources to perform specific actions, such as collecting dialed digits. *Note:* An FC is unique to the intelligent-network-IN/2 architecture.

functional profile. A standardization document that characterizes the requirements of a standard or group of standards, and specifies how the options and ambiguities in the standard(s) should be interpreted or implemented to (a) provide a particular information technology function, (b) provide for the development of uniform, recognized tests, and (c) promote interoperability among different network elements and terminal equipment that implement a specific profile.

functional signaling. In an integrated services digital network (ISDN), signaling in which the signaling messages are unambiguous and have clearly defined meanings that are known to both the sender and receiver of the messages. *Note:* Functional signaling is usually generated by the data terminal equipment (DTE).

functional signaling link. A combination of a communications link and the associated transfer control functions.

functional unit. An entity of hardware, software, or both, capable of accomplishing a specified purpose.

function signal. A set of signal elements that is used to transmit or represent a function-control character that actuates a control function, such as carriage return, line-feed, letters shift, or figures shift, that is to be performed by communications devices, such as teletypewriters and teleprinters. [From Weik '89]

fundamental. Of a periodic wave, the sinusoidal component, *i.e.,* Fourier component, having the lowest frequency. *Note:* Every periodic waveform may be expressed as the summation of the fundamental and its harmonics. For example, a square wave may be expressed as the summation of sine waves equal in frequency to the fundamental and all odd harmonics, each frequency having an appropriate amplitude and phase. A pure sinusoidal wave has only one component, *i.e.,* the fundamental. *Contrast with* harmonic, overtone.

fundamental mode. The lowest order mode of a waveguide. *Note:* In optical fibers, the fundamental mode is designated LP_{01} or HE_{11}.

furcate. *Synonym* break out.

fuse. 1. A device that has as its critical component a metal wire or strip that will melt when heated by a prescribed (design) amperage, creating an open in the circuit of which it is a part, thereby protecting the circuit from an overcurrent condition. *Note:* Fuses are often characterized as "fast-blow" or "slow-blow," according to the time required for them to respond to an overcurrent condition. Fast-blow fuses open nearly instantaneously when exposed to an overcurrent condition. Slow-blow fuses can tolerate a transient overcurrent condition, but will open if the overcurrent condition is sustained. 2. In optical fiber technology, to join the endfaces of a pair of optical fibers by melting, *i.e.*, welding, the endfaces together.

fused silica. *Synonym* vitreous silica.

fusion splice. In fiber optics, a splice created by localized heating of the ends of the two fibers to be joined. *Note:* A properly made fusion splice results in a continuous length of material with minimal discontinuities at the splice.

FWHM. *Abbreviation for* full width at half maximum.

FX. *Abbreviation for* fixed service, foreign exchange service.

G

gain. The ratio of output current, voltage, or power to input current, voltage, or power, respectively. *Note 1:* Gain is usually expressed in dB. *Note 2:* If the ratio is less than unity, the gain, expressed in dB, will be negative, in which case there is a loss between input and output.

gain hit. *See* hit.

gain medium. An active medium, device, or system in which amplification of input occurs with or without feedback. *Note:* Gain media include amplifiers, lasers, and avalanche photodiodes.

gain of an antenna. *Synonym* antenna gain.

galactic radio noise. *Synonym* cosmic noise.

gap loss. 1. The power loss that occurs when an optical signal is transferred from one fiber to another that is axially aligned with it, but longitudinally separated from it. *Note:* The gap allows light from the "transmitting" fiber to spread out as it leaves the fiber endface. When it strikes the "receiving" fiber, some of the light will enter the cladding, where it is quickly lost. 2. An analogous form of coupling loss that occurs between an optical source, *e.g.*, an LED, and an optical fiber. *Note:* Gap loss is not usually significant at the optical detector, because the sensitive area of the detector is normally somewhat larger than the cross section of the fiber core. Unless the separation is substantial, all light emerging from the fiber, even though it diverges, will still strike the detector. *Synonym* longitudinal offset loss.

gap-loss attenuator. An optical attenuator that exploits the principle of gap loss to reduce the optical power level when inserted in-line in the fiber path; *e.g.*, to prevent saturation of the receiver. *Note:* Gap-loss attenuators should be used in-line near the optical transmitter.

garble. 1. An error in transmission, reception, encryption, or decryption that changes the text of a message or any portion thereof in such a manner that it is incorrect or undecryptable. 2. In a telephone circuit or channel, readily audible but unintelligible interference from another circuit or channel. *Note:* Garble may, for example, take place in an FDM telephone carrier system in which an interfering signal from another channel or system is demodulated in such a fashion that it has an objectionable audio power level but is nonetheless unintelligible.

gate. 1. A device having one output channel and one or more input channels, such that the output channel state is completely determined by the input channel states, except during switching transients. 2. One of many types of combinational logic elements having at least two inputs; *e.g.*, AND, OR, NAND, and NOR.

gateway. 1. In a communications network, a network node equipped for interfacing with another network that uses different protocols. *Note 1:* A gateway may contain devices such as protocol translators, impedance matching devices, rate converters, fault isolators, or signal translators as necessary to provide system interoperability. It also requires that mutually acceptable administrative procedures be established between the two networks. *Note 2:* A protocol translation/mapping gateway interconnects networks with different network protocol technologies by performing the required protocol conversions. 2. *Loosely,* a computer configured to perform the tasks of a gateway.

gating. 1. The process of selecting only those portions of a wave between specified time intervals or between specified amplitude limits. 2. The controlling of signals by means of combinational logic elements. 3. A process in which a predetermined set of conditions, when established, permits a second process to occur.

gaussian beam. A beam of light whose electric field intensity distribution is gaussian. *Note:* When such a beam is circular in cross section the intensity at distance r from the center, $E(r)$, is given by

$$E(r) = E(0)\, e^{(-r/w)^2} ,$$

where $E(0)$ is the electrical field strength at the beam center, *i.e.*, at $r = 0$; and w is the value of r at which the intensity is 1/e of its value on the axis.

gaussian pulse. A pulse that has a waveform described by the gaussian distribution. *Note:* In the time domain, the amplitude of the waveform is given by

$$f(t) = A\, e^{(-t/\sigma)^2} ,$$

where A is the maximum amplitude, and σ is the pulse half-duration at the 1/e points.

GBH. *Abbreviation for* group busy hour.

GCT. *Abbreviation for* Greenwich Civil Time. *See* Coordinated Universal Time.

GDF. *Abbreviation for* group distribution frame.

gel. 1. A substance, resembling petroleum jelly in viscosity, that surrounds a fiber, or multiple fibers, enclosed in a loose buffer tube. *Note:* This gel serves to lubricate and support the fibers in the buffer tube. It also prevents water intrusion in the event the buffer tube is breached. 2. Index-matching material in the form of a gel. *Synonym* index-matching gel. *See* index-matching material.

general purpose computer. A computer designed to perform, or that is capable of performing, in a reasonably efficient manner, the functions required by both scientific and business applications. *Note:* A general purpose computer is often understood to be a large system, capable of supporting remote terminal operations, but it may also be a smaller computer, *e.g.*, a desktop workstation.

general purpose network. *See* common user network.

geometric optics. The branch of optics that describes light propagation in terms of rays. *Note 1:* Rays are bent at the interface between two dissimilar media, and may be curved in a medium in which the refractive index is a function of position. *Note 2:* The ray in geometric optics is perpendicular to the wavefront in physical optics. *Synonym* ray optics.

geometric spreading. *See* inverse-square law.

geostationary orbit. A circular orbit in the equatorial plane, any point on which revolves about the Earth in the same direction and with the same period as the Earth's rotation. *Note:* An object in a geostationary orbit will remain directly above a fixed point on the equator at a distance of approximately 42,164 km from the center of the Earth, *i.e.*, approximately 35,787 km above mean sea level.

geostationary satellite. A geosynchronous satellite whose circular and direct orbit lies in the plane of the Earth's equator and which thus remains fixed relative to the Earth; by extension, a satellite that remains approximately fixed relative to the Earth.

geostationary satellite orbit. The orbit in which a satellite must be placed to be a geostationary satellite.

geosynchronous orbit. Any orbit about the Earth, which orbit has a period equal to the period of rotation of the Earth about its axis, and in the same sense, *i.e.*, direction, as the rotation of the Earth.

germanium photodiode. A germanium-based PN- or PIN-junction photodiode. *Note 1:* Germanium photodiodes are useful for direct detection of optical wavelengths from approximately 1 μm to several tens of μm. *Note 2:* Germanium-based detectors are noisier than silicon-based detectors. Silicon-based detectors are therefore usually preferred for wavelengths shorter than 1 μm.

ghost. A secondary image or signal resulting from echo, envelope delay distortion, or multipath reception.

gigaflop. A billion, *i.e.*, 10^9, floating point operations per second.

gigahertz (GHz). A unit of frequency denoting 10^9 Hz.

glare. *Deprecated synonym for* call collision.

glass. 1. In the strict sense, a state of matter. 2. In fiber-optic communication, any of a number of noncrystalline, amorphous inorganic substances, formed, by heating, from metallic or semiconductor oxides or halides, and used as the material for fibers.

Note: The most common glasses are based on silicon dioxide (SiO_2).

glide slope facility. In aeronautical navigation, an instrument approach landing facility that furnishes vertical guidance information to an aircraft from its approach altitude down to the surface of the runway.

global. 1. Pertaining to, or involving, the entire world. 2. Pertaining to that which is defined in one subsection of an entity and used in at least one other subsection of the same entity. 3. In computer, data processing, and communications systems, pertaining to what is applicable to an area beyond the immediate area of consideration. *Note:* Examples of global entities are (a) in computer programming, an entity that is defined in one subdivision of a computer program and used in at least one other subdivision of that program and (b) in personal computer systems and their software packages, a setting, definition, or condition that applies to the entire software system.

global address. In a communications network, the predefined address that is used as an address for all users of that network, and that may not be the address of an individual user, or subgroup of users, of the network.

global status. 1. The set of attributes of an entity, described at a particular time, when that set is extended to every occurrence of that entity within a prescribed boundary. 2. The complete set of attributes necessary to describe an entity at a particular time.

GMT. *Abbreviation for* Greenwich Mean Time. *Obsolete term. See* Coordinated Universal Time.

go-ahead message. *Synonym* go-ahead notice.

go-ahead notice. In a tape-relay communications system, a service message, usually sent to a relay station or to a tributary station, that contains a request to the operator to resume transmitting over a specified channel or channels. *Synonyms* go-ahead message, start message, start notice.

go-ahead tone. In communications systems, an audible signal transmitted by a system indicating that the system is ready to receive a message or signal.

gold code. In spread-spectrum systems, a code that is generated by summing, using modulo-two addition, the outputs of two spread-spectrum code-sequence generators.

Gopher. A menu-based information searching tool that allows users to access various types of databases, such as FTP archives and white pages databases. *Note 1:* Gopher is most often used as an Internet browser. *Note 2:* Gopher software uses the client-server model.

GOS. *Abbreviation for* grade of service.

GOSIP. *Acronym for* Government Open Systems Interconnection Profile. A definition of Federal Government functional requirements for open systems computer network products, including a common set of Open System Interconnection (OSI) data communication protocols that enables systems developed by different vendors to interoperate and enable the users of different applications on these systems to exchange information. *Note 1:* The OSI protocols were developed primarily by ISO and CCITT. *Note 2:* The GOSIP is a subset of the OSI protocols and is based on agreements reached by vendors and users of computer networks participating in the National Institute of Standards and Technology (NIST) Implementors Workshop. *Note 3:* The GOSIP is described in the latest version of FIPS PUB 146.

Government Open Systems Interconnection Profile. *See* GOSIP.

graceful degradation. Degradation of a system in such a manner that it continues to operate, but provides a reduced level of service rather than failing completely.

graded-index fiber. An optical fiber with a core having a refractive index that decreases with increasing radial distance from the fiber axis. *Note:* The most common refractive index profile for a graded-index fiber is very nearly parabolic. The parabolic profile results in continual refocusing of the rays in the core, and compensates for multimode distortion.

graded-index profile. In the core of an optical fiber, a plot of the variation of refractive index such that the refractive index decreases with increasing radial distance from the fiber axis.

grade of service (GOS). 1. The probability of a call's being blocked or delayed more than a specified interval, expressed as a decimal fraction. *Note:* Grade of service may be applied to the busy hour or to some other specified period or set of traffic conditions. Grade of service may be viewed independently from the perspective of incoming versus outgoing calls, and is not necessarily equal in each direction. 2. In telephony, the quality of service for which a circuit is designed or conditioned to provide, *e.g.,* voice grade or program grade. *Note:* Criteria for different grades of service may include equalization for amplitude over a specified band of frequencies, or in the case of digital data transported via analog circuits, equalization for phase also.

grandfathered systems. Systems, including but not limited to, (a) PBX and key telephone systems, directly connected to the public switched telephone network on June 1, 1978, that may remain permanently connected thereto without registration unless subsequently modified, and (b) systems that are of the same type as those connected to the public switched telephone network on July 1, 1978, that were added before January 1, 1980, and that may remain permanently connected thereto without registration unless subsequently modified.

grandfathered terminal equipment. Terminal equipment (other than PBX and key telephone systems) and protective circuitry connected to the public switched telephone network before July 1, 1978, that may remain connected thereto for life without registration unless subsequently modified.

graphical user interface (GUI). A computer program or environment that displays options on the screen as icons, *i.e.,* picture symbols, by which users enter commands by selecting an icon. *Note:* Icons may be selected, *e.g.*, by pressing the <ENTER> key on the keyboard, by "clicking" a computer mouse button, or by touching the icon on a touch pad.

graphic character. 1. A visual representation of a character, other than a control character. 2. In the ASCII code, a character other than an alphanumeric character, intended to be written, printed, or otherwise displayed in a form that can be read by humans. *Note 1:* Graphic characters are contained in rows 2 through 7 of the ASCII code table. *Note 2:* The space and delete characters are considered to be graphic characters.

graphics. The art or science of conveying information through the use of display media, such as graphs, letters, lines, drawings, and pictures. *Note:* Graphics includes the transmission of coded images such as facsimile.

Gray code. A binary code in which consecutive decimal numbers are represented by binary expressions that differ in the state of one, and only one, one bit. *Synonym* reflected code.

gray scale. An optical pattern consisting of discrete steps or shades of gray between black and white.

great circle. A circle defined by the intersection of the surface of the Earth and any plane that passes through the center of the Earth. *Note:* On the idealized surface of the Earth, the shortest distance between two points lies along a great circle.

Greenwich Civil Time (GCT). *Synonym* Greenwich Mean Time (GMT). *Obsolete term. See* Coordinated Universal Time.

Greenwich Mean Time (GMT). Mean solar time at the meridian of Greenwich, England, formerly used as a basis for standard time throughout the world. *Obsolete term. Synonym* Greenwich Civil Time. *See* Coordinated Universal Time.

ground. 1. An electrical connection to earth through an earth-electrode subsystem. 2. In an electrical circuit, a common return path that usually (a) is connected to an earth-electrode subsystem and (b) is extended throughout a facility via a facility ground system consisting of the signal reference subsystem, the fault protection subsystem, and the lightning protection subsystem. 3. In an electrical circuit, a common return path that (a) may not necessarily be connected to earth and (b) is the zero voltage reference level for the equipment or system.

ground absorption. The dissipation of rf energy by the Earth.

ground constants. The electrical parameters of earth, such as conductivity, permittivity, and magnetic permeability. *Note 1:* The values of these parameters vary with the local chemical composition and density of the earth. *Note 2:* For a propagating electromagnetic wave, such as a surface wave propagating along the surface of the Earth, these parameters vary with frequency and direction.

ground current. In the presence of an electrical fault, the current that flows in the protective ground wire of a power distribution system. *Contrast with* ground loop.

ground loop. In an electrical system, an unwanted current that flows in a conductor connecting two points that are nominally at the same potential, *i.e.,* ground, but are actually at different potentials. *Note 1:* For example, the electrical potential at different points on the surface of the Earth can vary by hundreds of volts, primarily from the influence of the solar wind. Such an occurrence can be hazardous, *e.g.,* to personnel working on long grounded conductors such as metallic telecommunications cable pairs. *Note 2:* A ground loop can also exist in a floating ground system, *i.e.,* one not connected to an Earth ground, if the conductors that constitute the ground system have a relatively high resistance, or have, flowing through them, high currents that produce a significant voltage ("I•R") drop. *Note 3:* Ground loops can be detrimental to the operation of the electrical system. *Contrast with* ground current.

ground plane. An electrically conductive surface that serves as the near-field reflection point for an antenna. *Note:* A ground plane may consist of a natural (*e.g.,* Earth or sea) surface, an artificial surface of opportunity (*e.g.,* the roof of a motor vehicle), or a specially designed artificial surface (*e.g.,* the disc of a discone antenna).

ground potential. The zero reference level used to apply and measure voltages in a system. *Note:* A potential difference may exist between this reference level and the ground potential of the Earth, which varies with locality, soil conditions, and meteorological phenomena.

ground-return circuit. 1. A circuit using a common return path that is at ground potential. *Note:* Earth may serve as a portion of the ground-return circuit. 2. A circuit in which there is a common return path, whether or not connected to earth.

ground start. A method of signaling from a terminal or subscriber loop to a switch, in which method one side of a cable pair is temporarily grounded.

ground wave. In radio transmission, a surface wave that propagates close to the surface of the Earth. *Note 1:* The Earth has one refractive index and the atmosphere has another, thus constituting an interface that supports surface wave transmission. These refractive indices are subject to spatial and temporal changes. *Note 2:* Ground waves do not include ionospheric and tropospheric waves.

group. 1. In frequency-division multiplexing, a specific number of associated voice channels, either within a supergroup or as an independent entity. *Note 1:* In wideband systems, a group usually consists of 12 voice channels and occupies the frequency band from 60 kHz to 108 kHz. *Note 2:* this is CCITT group B. *Note 3*: CCITT Basic Group A, for carrier telephone systems, consists of 12 channels occupying upper sidebands in the 12-kHz to 60-kHz band. Basic Group A is no longer mentioned in CCITT Recommendations. *Note 4:* A supergroup usually consists of 60 voice channels, *i.e.,* 5 groups of 12 voice channels each, occupying the frequency band from 312 kHz to 552 kHz. *Note 5:* A mastergroup consists of 10 supergroups or 600 voice channels. *Note 6:* The CCITT standard mastergroup consists of 5 supergroups. The U.S. commercial carrier standard mastergroup consists of 10 supergroups. *Note 7:* The terms "supermaster group" or "jumbo group" are sometimes used to refer to 6 mastergroups. 2. A set of characters forming a unit for transmission or cryptographic treatment.

group address. In a communications network, a predefined address used to address only a specified set of users. *Synonym* collective address.

group alerting and dispatching system. A service feature that (a) enables a controlling telephone to place a call to a specified number of telephones simultaneously, (b) enables the call to be recorded, (c) if any of the called lines is busy, enables the equipment to camp on until the busy line is free, and (d) rings the free line and plays the recorded message.

group busy hour (GBH). The busy hour for a given trunk group.

group delay. 1. The rate of change of the total phase shift with respect to angular frequency, $d\theta/d\omega$, through a device or transmission medium, where θ is the total phase shift, and ω is the angular frequency equal to $2\pi f$, where f is the frequency. 2. In an optical fiber, the transit time required for optical power, traveling at a given mode's group velocity, to travel a given distance. *Note:* For optical fiber dispersion measurement purposes, the quantity of interest is group delay per unit length, which is the reciprocal of the group velocity of a particular mode. The measured group delay of a signal through an optical fiber exhibits a wavelength dependence due to the various dispersion mechanisms present in the fiber.

group delay time. In a group of waves that have slightly different individual frequencies, the time required for any defined point on the envelope (*i.e.,* the envelope determined by the additive resultant of the group of waves) to travel through a device or transmission facility.

group distribution frame (GDF). In frequency-division multiplexing, a distribution frame that provides terminating and interconnecting facilities at the group level, *i.e.*, group modulator output and group demodulator input circuits of FDM carrier equipment. *Note:* The basic spectrum of the FDM group is 60 kHz to 108 kHz.

***group index (*N*).** In fiber optics, for a given mode propagating in a medium of refractive index η, the velocity of light in vacuum, c, divided by the group velocity of the mode. *Note:* For a plane wave of wavelength λ, the group index may also be expressed,

$$N = n - \lambda \frac{dn}{d\lambda},$$

where *n* is the phase index of wavelength λ.

grouping factor. *Synonym* blocking factor.

group 1. . . 4 facsimile. *See* facsimile.

group patch bay. *See* patch bay.

group velocity. 1. The velocity of propagation of an envelope produced when an electromagnetic wave is modulated by, or mixed with, other waves of different frequencies. *Note:* The group velocity is the velocity of information propagation and, loosely, of energy propagation. 2. In optical fiber transmission, for a particular mode, the reciprocal of the rate of change of the phase constant with respect to angular frequency. *Note:* The group velocity equals the phase velocity if the phase constant is a linear function of the angular frequency, $\omega = 2\pi f$, where *f* is the frequency. 3. In optical-fiber transmission, the velocity of the modulated optical power.

G/T. *Abbreviation for* antenna gain-to-noise-temperature.

guard band. *See* frequency guard band, time guard band.

guarded frequency. A transmission frequency that is not to be jammed or interfered with because of the value of the information being derived from it. *Note:* For example, a guarded frequency will not be jammed when the tactical, strategic, and technical information that can be obtained from the transmissions outweighs the potential operational gain achieved by jamming.

guided mode. *Synonym* bound mode.

guided ray. In an optical fiber, a ray that is confined primarily to the core. *Note:* A guided ray satisfies the relation given by

$$0 \leq \sin \theta_r \leq \sqrt{n_r^2 - n_a^2} ,$$

where θ_r is the angle the ray makes with the fiber axis, r is the radial position, *i.e.*, radial distance, of the ray from the fiber axis, n_r is the refractive index at the radial distance r from the fiber axis, and n_a is the refractive index at the core radius, a, *i.e.*, at the core-cladding interface. Guided rays correspond to bound modes, *i.e.*, guided modes, in terms of modes rather than rays. *Synonyms* bound ray, trapped ray.

guided wave. A wave having (a) energy concentrated near a boundary, or between substantially parallel boundaries, separating materials of different properties and (b) a direction of propagation effectively parallel to these boundaries.

Hagelbarger code. A convolutional code that enables error bursts to be corrected provided that there are relatively long error-free intervals between the error bursts. *Note:* In the Hagelbarger code, inserted parity check bits are spread out in time so that an error burst is not likely to affect more than one of the groups in which parity is checked.

half-duplex (HDX) operation. Operation in which communication between two terminals occurs in either direction, but in only one direction at a time. *Note:* Half-duplex operation may occur on a half-duplex circuit or on a duplex circuit, but it may not occur on a simplex circuit. *Synonyms* one-way reversible operation, two-way alternate operation.

halftone. Any photomechanical printing surface or the impression therefrom in which detail and tone values are represented by a series of evenly spaced dots in varying size and shape, varying in direct proportion to the intensity of tones they represent.

halftone characteristic. 1. In facsimile systems, the relationship between the density of the recorded copy and the density of the object, *i.e.*, the original. 2. In facsimile systems, the relationship between the amplitude of the facsimile signal to either the density of the object or the density of the recorded copy when only a portion of the system is under consideration. *Note:* In an FM facsimile system, an appropriate parameter other than the amplitude is used.

Hamming code. An error-detecting and error-correcting binary code, used in data transmission, that can (a) detect all single- and double-bit errors and (b) correct all single-bit errors. *Note:* A Hamming code satisfies the relation $2^m \geq n+1$, where n is the total number of bits in the block, k is the number of information bits in the block, and m is the number of check bits in the block, where $m = n - k$.

Hamming distance. The number of digit positions in which the corresponding digits of two binary words of the same length are different. *Note 1:* The Hamming distance between 1011101 and 1001001 is two. *Note 2:* The concept can be extended to other notation systems. For example, the Hamming distance between 2143896 and 2233796 is three, and between "toned" and "roses" it is also three. *Synonym* signal distance.

Hamming weight. The number of non-zero symbols in a symbol sequence. *Note:* For binary signaling, Hamming weight is the number of "1" bits in the binary sequence.

handoff. 1. In cellular mobile systems, the process of transferring a phone call in progress from one cell transmitter and receiver and frequency pair to another cell transmitter and receiver using a different frequency pair without interruption of the call. 2. In satellite communications, the process of transferring ground-station control responsibility from one ground station to another without loss or interruption of service.

handshaking. 1. In data communications, a sequence of events governed by hardware or software, requiring mutual agreement of the state of the operational modes prior to information exchange. 2. The process used to establish communications parameters between two stations. *Note:* Handshaking follows the establishment of a circuit between the stations and precedes information transfer. It is used to agree upon such parameters as information transfer rate, alphabet, parity, interrupt procedure, and other protocol features.

hangover. *Synonym* tailing.

HA1-receiver weighting. A noise weighting used in a noise measuring set to measure noise across the HA1-receiver of a 302-type or similar instrument. *Note 1:* The meter scale readings of an HA1 test set are in dBa (HA1). *Note 2:* HA1 noise weighting is obsolete for new DOD applications.

hard copy. In computer graphics and in telecommunications, a permanent reproduction, on any media suitable for direct use by a person, of displayed or transmitted data. *Note 1:* Examples of hard copy include teletypewriter pages, continuous printed tapes, facsimile pages, computer printouts, and radiophoto prints. *Note 2:* Magnetic tapes, diskettes, and nonprinted punched paper tapes are not hard copy.

hard disk. A flat, circular, rigid plate with a magnetizable surface on one or both sides of which data can be stored. *Note:* A hard disk is distinguished from a diskette by virtue of the fact that it is rigid. Early in the development of computer technology, hard disks, often multiple disks mounted on a common spindle, were interchangeable and removable from their drives, which were separate from the processor chassis. This technology is still in use, especially in conjunction with large mainframe computers, but physically smaller computers use hard disks that are in sealed units, along with their control electronics and read/write heads. The sealed units are usually installed permanently in the same chassis that contains the processor.

hardened. Pertaining to the condition of a facility with protective features that enable it to withstand destructive forces, such as explosions, natural disasters, or ionizing radiation.

hard limiting. *See* limiting.

hard sectoring. In magnetic or optical disk storage, sectoring that uses a physical mark on the disk, from which mark sector locations are referenced. *Note:* Hard sectoring may be done, for example, by punching an index hole in a floppy diskette. When the presence of the index hole is recognized by an optical reader, a reference signal is generated. All sector locations can be referenced from this signal.

hardware. 1. Physical equipment as opposed to programs, procedures, rules, and associated documentation. 2. The generic term dealing with physical items as distinguished from its capability or function such as equipment, tools, implements, instruments, devices, sets, fittings, trimmings, assemblies, subassemblies, components, and parts. *Note: Hardware* is often used in regard to the stage of development, as in the passage of a device or component from the design stage into the hardware stage as the finished object. 3. In data automation, the physical equipment or devices forming a computer and peripheral components.

hardwire. 1. To connect equipment or components permanently in contrast to using switches, plugs, or connectors. 2. To wire in fixed logic or read-only storage that cannot be altered by program changes.

harmful interference. 1. Any emission, radiation, or induction interference that endangers the functioning or seriously degrades, obstructs, or repeatedly interrupts a communications system, such as a radio navigation service, telecommunications service, radio communications service, search and rescue service, or weather service, operating in accordance with approved standards, regulations, and procedures. *Note:* To be considered harmful interference, the interference must cause serious detrimental effects, such as circuit outages and message losses, as opposed to interference that is merely a nuisance or annoyance that can be overcome by appropriate measures. 2. Interference which endangers the functioning of a radionavigation service or of other safety services or seriously degrades, obstructs, or repeatedly interrupts a radiocommunication service operating in accordance with these *[Radio] Regulations.*

harmonic. 1. Of a sinusoidal wave, an integral multiple of the frequency of the wave. *Note:* The frequency of the sine wave is called the fundamental frequency or the first harmonic, the second harmonic is twice the fundamental frequency, the third harmonic is thrice the fundamental frequency, *etc.* 2. Of a periodic signal or other periodic phenomenon, such as an electromagnetic wave or a sound wave, a component frequency of the signal that is an integral multiple of the fundamental frequency. *Note:* The fundamental frequency is the reciprocal of the period of the periodic phenomenon. *Contrast with* fundamental, overtone.

harmonic distortion. In the output signal of a device, distortion caused by the presence of frequencies that are not present in the input signal. *Note:* Harmonic distortion is caused by nonlinearities within the device.

hazards of electromagnetic radiation to fuel (HERF). The potential for electromagnetic radiation to cause ignition or detonation of volatile combustibles, such as aircraft fuels.

hazards of electromagnetic radiation to ordnance (HERO). The potential for electromagnetic radiation to affect adversely munitions or electroexplosive devices.

hazards of electromagnetic radiation to personnel (HERP). The potential for electromagnetic radiation to produce harmful biological effects in humans.

H-bend. A smooth change in the direction of the axis of a waveguide, throughout which the axis remains in a plane parallel to the direction of magnetic H-field (transverse) polarization. *Synonym* H-plane bend.

H-channel. In Integrated Services Digital Networks (ISDN), a 384-kb/s, 1472-kb/s, or 1536-kb/s channel, designated as "H_0", "H_{10}", and "H_{11}", respectively, accompanied by timing signals used to carry a wide variety of user information. *Note:* Examples of types of user information representation forms include fast facsimile, video, high-speed data, high-quality audio, packet-switched data, bit streams at rates less than the respective H-channel bit rate that have been rate-adapted or multiplexed together, and packet-switched information.

HDLC. *Abbreviation for* high-level data link control.

HDTV. *Abbreviation for* high-definition television.

HDX. *Abbreviation for* half duplex.

head. A device that reads, writes, and/or erases data on a storage medium.

head end. 1. A central control device required by some networks (*e.g.,* LANs or MANs) to provide such centralized functions as remodulation, retiming, message accountability, contention control, diagnostic control, and access to a gateway. 2. A central control device, within CATV systems, that provides centralized functions such as remodulation.

header. The portion of a message that contains information used to guide the message to the correct destination. *Note:* Examples of items that may be in a header are the addresses of the sender and receiver, precedence level, routing instructions, and synchronizing bits.

head-of-bus function. The function that generates management information and empty bus slots at the point on each bus where data flow begins.

head-on collision. A collision that occurs on a communications channel when two or more users begin to transmit on the channel at approximately the same instant.

Heaviside layer. *Synonym* E region.

height gain. For a given propagation mode of an electromagnetic wave, the ratio of the field strength at a specified height to the field strength at the surface of the Earth.

helical antenna. An antenna that has the form of a helix. *Note:* When the helix circumference is much smaller than one wavelength, the antenna radiates at right angles to the axis of the helix. When the helix circumference is one wavelength, maximum radiation is along the helix axis.

HEMP. *Abbreviation for* high-altitude electromagnetic pulse.

HE_{11} mode. Designation for the fundamental hybrid mode of an optical fiber.

HERF. *Abbreviation for* hazards of electromagnetic radiation to fuel.

HERO. *Abbreviation for* hazards of electromagnetic radiation to ordnance.

HERP. *Abbreviation for* hazards of electromagnetic radiation to personnel.

hertz (Hz). 1. The SI unit of frequency, equal to one cycle per second. *Note:* A periodic phenomenon that has a period of one second has a frequency of one hertz. 2. A unit of frequency which is equivalent to one cycle per second.

Hertzian wave. *Synonym* radio wave.

heterochronous. A relationship between two signals such that their corresponding significant instants do not necessarily occur at the same time. *Note:* Two signals having different nominal signaling rates and not stemming from the same clock or from homochronous clocks are usually heterochronous.

heterodyne. 1. To generate new frequencies by mixing two or more signals in a nonlinear device such as a vacuum tube, transistor, or diode mixer. *Note:* A superheterodyne receiver converts any selected incoming frequency by heterodyne action to a common intermediate frequency where amplification and selectivity (filtering) are provided. 2. A frequency produced by mixing two or more signals in a nonlinear device.

heterodyne repeater. In radio reception and retransmission, a repeater that converts the original band of frequencies of the received signal to a different frequency band for retransmission after amplification. *Note:* Heterodyne repeaters are used, for example, in microwave systems, to avoid undesired feedback between the receiving and transmitting antennas. *Synonym* IF repeater.

heterogeneous multiplexing. Multiplexing in which not all the information-bearer channels operate at the same data signaling rate.

heuristic routing. Routing in which data, such as time delay, extracted from incoming messages, during specified periods and over different routes, are used to determine the optimum routing for transmitting data back to the sources. *Note:* Heuristic routing allows a measure of route optimization based on recent empirical knowledge of the state of the network.

hexadecimal. 1. Characterized by a selection, choice or condition that has sixteen possible different values or states. *Synonym* sexadecimal. 2. Pertaining to a fixed-radix numeration system in which the radix is sixteen.

HF. *Abbreviation for* high frequency.

HFDF. *Abbreviation for* high-frequency distribution frame.

hierarchical computer network. A computer network in which processing and control functions are performed at several levels by computers specially suited for the functions performed, such as industrial process control, inventory control, database control, or hospital automation.

hierarchically synchronized network. A mutually synchronized network in which some clocks exert more control than others, the network operating frequency being a weighted mean of the natural frequencies of the population of clocks.

hierarchical routing. Routing that is based on hierarchical addressing. *Note:* Most Transmission Control Protocol/Internet Protocol (TCP/IP) routing is based on a two-level hierarchical routing in which an IP address is divided into a network portion and a host portion. Gateways use only the network portion until an IP datagram reaches a gateway that can deliver it directly. Additional levels of hierarchical routing are introduced by the addition of subnetworks.

high-altitude electromagnetic pulse (HEMP). An electromagnetic pulse produced at an altitude effectively above the sensible atmosphere, *i.e.*, above about 120 km.

high-definition television (HDTV). Television that has approximately twice the horizontal and twice the vertical emitted resolution specified by the NTSC standard. *Note 1:* In HDTV, the total number of pixels is therefore approximately four times that of the NTSC standard. *Note 2:* HDTV may include any or all improved-definition television (IDTV) and extended-television (EDTV) improvements. *Note 3:* HDTV employs a wide aspect ratio.

higher frequency ground. *Deprecated term name. See* facility grounding system.

high frequency (HF). Frequencies from 3 MHz to 30 MHz.

high-frequency distribution frame (HFDF). A distribution frame that provides terminating and interconnecting facilities for those combined supergroup modulator output circuits and combined supergroup demodulator input circuits that contain signals occupying the baseband spectrum.

high-level control. In the hierarchical structure of a primary or secondary data transmission station, the conceptual level of control or processing logic that (a) is above the Link Level and (b) controls Link Level functions, such as device control, buffer allocation, and station management.

high-level data link control (HDLC). A Link-Level protocol used to facilitate reliable point-to-point transmission of a data packet. *Note:* A subset of HDLC, known as "*LAP-B,*" is the Layer-two protocol for CCITT Recommendation X.25.

high-level language (HLL). A computer programming language that is primarily designed for, and syntactically oriented to, particular classes of problems and that is essentially independent of the structure of a specific computer or class of computers; for example, Ada®, COBOL, Fortran, Pascal. *Synonym* high-order language.

high-order language. *Synonym* high-level language.

high-pass filter. A filter that passes frequencies above a given frequency and attenuates all others.

high-performance equipment. Equipment that (a) has the performance characteristics required for use in trunks or links, (b) is designed primarily for use in global and tactical systems, and (c) sufficiently withstands electromagnetic interference when operating in a variety of network or point-to-point circuits. *Note:* Requirements for global and tactical high-performance equipment may differ.

high-usage trunk group. A group of trunks for which an alternate route has been provided to absorb the relatively high rate of overflow traffic.

highway. 1. A digital serial-coded bit stream with time slots allotted to each call on a sequential basis. 2. A common path or a set of parallel paths over which signals from more than one channel pass with separation achieved by time division.

hiss. Noise in the audio frequency range, having subjective characteristics analogous to prolonged sibilant sounds. *Note:* Noise in which there are no pronounced low-frequency components may be considered as hiss.

hit. 1. A transient disturbance to a communication medium. 2. A match of data to a prescribed criterion.

HLL. *Abbreviation for* high-level language.

hockey puck. A polishing fixture used to facilitate the manual finishing of the endfaces of certain types of optical fiber connectors. *Note 1:* The hockey puck consists of the appropriate mating sleeve for the connector in question, mounted at right angles to, and in the center of, a disk of stainless steel or other hard material. When the unfinished connector, secured to the fiber-optic cable, is mounted in the hockey puck, excess material (*e.g.*, fiber end, bead of adhesive material, and excess connector length, if present) protrudes from the opposite side of the disk. The excess is then ground away as the fixture is manually swept to and fro, usually in a figure-8 pattern, in contact with a piece of microfinishing film which is in turn supported by a rigid flat substrate. Two to four grades of microfinishing film, with abrasive particles ranging in size from 15 μm to 0.3 μm, are commonly used. *Note 2:* Various manufacturers use proprietary names to identify this device; however "hockey puck" has become ubiquitous.

hold-in frequency range. The range of frequencies over which a phase-locked loop can vary and still maintain frequency lock.

holding time. The total length of time that a call makes use of a trunk or channel. *Note:* Holding time is usually measured in seconds.

homing. 1. A process in which a mobile station is directed, or directs itself, toward an electromagnetic, thermal, sonic, or other source of energy, whether primary or reflected, or follows a vector force field or a gradient of a scalar force field. 2. In radio direction finding, the locating of a moving signal source by a moving direction-finding station that has a mobile advantage. 3. The act of approaching a source of electromagnetic radiation in which the approaching vehicle is guided by a receiver with a directional antenna. 4. Seeking, finding, intercepting, and engaging an object, *i.e.*, a target (fixed or mobile) that may contain a signal source.

homochronous. The relationship between two signals such that their corresponding significant instants are displaced by a constant interval of time.

homogeneous cladding. In an optical fiber, a cladding in which the nominal refractive index is constant throughout. *Note:* An optical fiber may have several homogeneous claddings, each having a different refractive index.

homogeneous multiplexing. Multiplexing in which all of the information-bearer channels operate at the same data signaling rate.

hop. 1. The excursion of a radio wave from the Earth to the ionosphere and back to the Earth. *Note:* The number of hops indicates the number of reflections from the ionosphere. 2. A waveform transmitted for the duration of each relocation of the carrier frequency of a frequency-hopped system. 3. To modify a modulated waveform with constant center frequency so that it frequency hops.

hop count. 1. In a data communications network, the number of legs traversed by a packet between its source and destination. *Note:* Hop count may be used to determine the time-to-live for some packets. 2. The number of signal regenerating devices (such as repeaters, bridges, routers, and gateways) through which data must pass to reach their destination.

horizon angle. Of a directional antenna, the angle, in a vertical plane, subtended by the lines extending (a) from the antenna to the radio horizon and (b) from the antenna in its direction of maximum radiation.

horizontal redundancy check. *Synonym* longitudinal redundancy check.

horizontal resolution. In facsimile, the number of picture elements per unit distance in the direction of scanning or recording.

horn. 1. In radio transmission, an open-ended waveguide, of increasing cross-sectional area, which radiates directly in a desired direction or feeds a reflector that forms a desired beam. *Note 1:* Horns may have one or more expansion curves, *i.e.*, longitudinal cross sections, such as elliptical, conical, hyperbolic, or parabolic curves, and not necessarily the same expansion curve in each (E-plane and H-plane) cross section. *Note 2:* A very wide range of beam patterns may be formed by controlling horn dimensions and shapes, placement of the reflector, and reflector shape and dimensions. 2. A portion of a waveguide in which the cross section is smoothly increased along the axial direction. 3. In audio systems, a tube, usually having a rectangular transverse cross section and a linearly or exponentially increasing cross-sectional area, used for radiating or receiving acoustic waves.

horn gap switch. A switch provided with arcing horns, ordinarily used for disconnecting or breaking the charging current of overhead transmission and distribution lines.

host. 1. In packet- and message-switching communications networks, the collection of hardware and software that makes use of packet or message switching to support user-to-user, *i.e.,* end-to-end, communications, interprocess communications, and distributed data processing. 2. *Synonym* host computer.

host computer. 1. In a computer network, a computer that provides end users with services such as computation and database access and that usually performs network control functions. *Synonym* host. 2. A computer on which is developed software intended to be used on another computer.

hot boot. *Synonym* warm restart.

hotline. A point-to-point communications link in which a call is automatically originated to the preselected destination without any additional action by the user when the end instrument goes off-hook. *Note 1:* Hotlines cannot be used to originate calls other than to preselected destinations. *Note 2:* Various priority services that require dialing are *not* properly termed "hotlines." *Synonyms* automatic signaling service, off-hook service.

hot standby. *See* standby.

house cable. *Deprecated term.* Communication cable within a building or a complex of buildings. *Note:* House cable owned before divestiture by the Bell System and after divestiture by the Regional Bell Operating Companies will eventually be fully depreciated and will then belong to the customer. *See* on-premises wiring.

housekeeping signals. *Synonym* service signals.

H-plane bend. *Synonym* H-bend.

HTML. *Abbreviation for* Hypertext Markup Language. An application of SGML (Standard Generalized Markup Language [ISO 8879]) implemented in conjunction with the World Wide Web to facilitate the electronic exchange and display of simple documents using the Internet.

HTTP. *Abbreviation for* Hypertext Transfer Protocol. In the World Wide Web, a protocol that facilitates the transfer of hypertext-based files between local and remote systems.

hub. 1. A distribution point in a network. 2. A device that accepts a signal from one point and redistributes it to one or more points.

Huffman coding. A coding technique used to compact data by representing the more common events with short codes and the less common events with longer codes. *Note:* Huffman coding is used in Group 3 facsimile.

hundred call-seconds (CCS). *See* call-second.

hunting. 1. In telephony, pertaining to the operation of a selector or other similar device to find and establish a connection with an idle circuit of a chosen group. 2. Pertaining to the failure of a device to achieve a state of equilibrium, usually by alternately overshooting and undershooting the point of equilibrium.

hybrid. 1. A functional unit in which two or more different technologies are combined to satisfy a given requirement. *Note:* Examples of hybrids include (a) an electronic circuit having both vacuum tubes and transistors, (b) a mixture of thin-film and discrete integrated circuits, and (c) a computer that has both analog and digital capability. 2. A resistance hybrid. 3. A hybrid coil.

hybrid balance. An expression of the degree of electrical symmetry between two impedances connected to two conjugate sides of a hybrid set or resistance hybrid. *Note 1:* Hybrid balance is usually expressed in dB. *Note 2:* If the respective impedances of the branches of the hybrid that are connected to the conjugate sides of the hybrid are known, hybrid balance may be computed by the formula for return loss.

hybrid cable. An optical communications cable having two or more different types of optical fibers, *e.g.*, single-mode and multimode fibers. *Contrast with* composite cable.

hybrid coil. A single transformer that effectively has three windings, and which is designed to be configured as a circuit having four branches, *i.e.,* ports, that are conjugate in pairs. *Note:* The primary use of a hybrid coil is to convert between 2-wire and 4-wire operation in concatenated sections of a communications circuit. Such conversion is necessary when repeaters are introduced in a 2-wire circuit. *Synonym* bridge transformer.

hybrid communications network. A communications network that uses a combination of line facilities, *i.e.,* trunks, loops, or links, some of which use only analog or quasi-analog signals and some of which use only digital signals. *Synonym* hybrid system.

hybrid computer. A computer that processes both analog and digital data.

hybrid connector. A connector that contains contacts for more than one type of service. *Note:* Examples of hybrid connectors are those that have contacts for both optical fibers and twisted pairs, electric power and twisted pairs, or shielded and unshielded twisted pairs.

hybrid coupler. In an antenna system, a hybrid junction used as a directional coupler. *Note:* The loss through a hybrid coupler is usually ≈3 dB.

hybrid interface structure. In integrated services digital networks (ISDN), an interface structure that uses both labeled and positioned channels.

hybrid junction. A waveguide or transmission line arranged such that (a) there are four ports, (b) each port is terminated in its characteristic impedance, and (c) energy entering any one port is transferred, usually equally, to two of the three remaining ports. *Note:* Hybrid junctions are used as mixing or dividing devices.

hybrid mode. A mode consisting of components of both electrical and magnetic field vectors in the direction of propagation. *Note:* In fiber optics, such modes correspond to skew (nonmeridional) rays.

hybrid network. *See* hybrid communications network.

hybrid routing. Routing in which numbering plans and routing tables are used to permit the collocation, in the same area code, of switches using a deterministic routing scheme with switches using a nondeterministic routing scheme, such as flood search routing. *Note:* Routing tables are constructed with no duplicate numbers, so that direct dial service can be provided to all network subscribers. This may require the use of 10-digit numbers.

hybrid set. Two or more transformers interconnected to form a network having four ports that are conjugate in pairs. *Note:* The primary use of a hybrid set is to convert between 2-wire and 4-wire operation in concatenated sections of a communications circuit. Such conversion is necessary when repeaters are introduced in a 2-wire circuit.

hybrid spread spectrum. A combination of frequency hopping spread spectrum and direct-sequence spread spectrum.

hybrid system. *Synonym* hybrid communications network.

hybrid topology. *See* network topology.

hydroxyl ion absorption. In optical fibers, the absorption of electromagnetic waves, including the near-infrared, due to the presence of trapped hydroxyl ions remaining from water as a contaminant. *Note:* The hydroxyl (OH^-) ion can penetrate glass during or after product fabrication, resulting in significant attenuation of discrete optical wavelengths, *e.g.*, approximately 1.3 μm, used for communications via optical fibers.

hypermedia. Computer-addressable files that contain pointers for linking to multimedia information, such as text, graphics, video, or audio in the same or other documents. *Note:* The use of hypertext links is known as navigating.

hypertext. The system of coding that is used to create or navigate hypermedia in a nonsequential manner.

Hypertext Transfer Protocol. *See* HTTP.

Hz. *Abbreviation for* hertz.

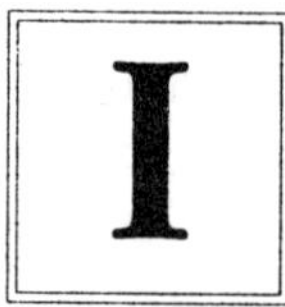

ICA. *Abbreviation for* International Communications Association.

ICI. *Abbreviation for* incoming call identification.

ICMP. *Abbreviation for* Internet Control Message Protocol. An Internet protocol that reports datagram delivery errors. *Note 1:* ICMP is a key part of the TCP/IP protocol suite. *Note 2:* The packet internet gopher (ping) application is based on ICMP.

icon. In computer systems, a small, pictorial representation of an application software package, idea, or concept used in a window or a menu to represent commands, files, or options.

ICW. *Abbreviation for* interrupted continuous wave.

identification, friend or foe (IFF). A system using electromagnetic transmissions to which equipment carried by friendly forces automatically responds, for example, by emitting pulses, thereby distinguishing themselves from enemy forces. *Note:* The secondary surveillance radar (SSR) system used in modern air traffic control systems is an outgrowth of the military IFF system used during World War II. The IFF equipment carried by modern military aircraft is compatible with the transponder system used for civilian air traffic control.

identification friend or foe personal identifier. The discrete identification, friend or foe code assigned to a particular aircraft, ship, or other vehicle for identification by electronic means.

identifier (ID). 1. In telecommunications and data processing systems, one or more characters used to identify, name, or characterize the nature, properties, or contents of a set of data elements. 2. A string of bits or characters that names an entity, such as a program, device, or system, in order that other entities can call that entity. 3. In programming languages, a lexical unit that names a language object, such as a variable, array, record, label, or procedure. *Note:* An identifier is placed in a label. The label is attached to, is a part of, or remains associated with, the information it identifies. If the label becomes disassociated from the information it identifies, the information may not be accessible.

IDF. *Abbreviation for* intermediate distribution frame.

idle-channel noise. Noise that is present in a communications channel when no signals are applied. *Note:* The channel conditions and terminations must be stated for idle-channel noise measurements to be meaningful.

idle character. A control character that is transmitted when no useful information is being transmitted.

idle-line termination. A switch-controlled electrical network that maintains a desired impedance at a trunk or line terminal that is in the idle state.

idle state. The telecommunications service condition that exists whenever user messages are not being transmitted but the service is immediately available for use.

idle time. A period during which a system, circuit, or component is not in use, but is available.

IDN. *Abbreviation for* integrated digital network.

IDTV. *Abbreviation for* improved-definition television.

IF. *Abbreviation for* intermediate frequency.

I/F. *Abbreviation for* interface.

IFF. *Abbreviation for* identification, friend or foe.

IFRB. *Abbreviation for* International Frequency Registration Board.

IF repeater. *See* heterodyne repeater.

IFS. *Abbreviation for* ionospheric forward scatter. *See* ionospheric scatter.

ILD. *Abbreviation for* injection laser diode.

illegal character. A character, or a combination of bits, that is not valid in a given system according to specified criteria, such as with respect to a specified alphabet, a particular pattern of bits, a rule of formation, or a check code. *Synonyms* false character, forbidden character, improper character, unallowable character, unused character.

ILS. *Abbreviation for* instrument landing system.

IM. *Abbreviation for* intensity modulation, intermodulation.

image. In the field of image processing, a two-dimensional representation of a scene. *Synonym* **picture.**

image antenna. A hypothetical mirror-image, *i.e.,* virtual-image, of an antenna, *i.e.,* antenna element, considered to extend as far below ground, *i.e.,* the ground plane, as the actual antenna is above the ground plane. *Note 1:* The image antenna is helpful in calculating electric field vectors, magnetic field vectors, and electromagnetic fields emanating from the real antenna, particularly in the vicinity of the antenna and along the ground. Each charge and current in the real antenna has its image that may also be considered as a source of radiation equal to, but differently directed from, its real counterpart. *Note 2:* An image antenna may also be considered to be on the opposite side of any equipotential plane surface, such as a metal plate acting as a ground plane, analogous to the position of a virtual optical image in a plane mirror. *Note 3:* The ground plane need not be grounded to the Earth.

image frequency. In radio reception using heterodyning in the tuning process, an undesired input frequency that is capable of producing the same intermediate frequency (IF) that the desired input frequency produces. *Note:* The term *image* arises from the mirror-like symmetry of signal and image frequencies about the beating-oscillator frequency.

image frequency rejection ratio. *Synonym* image rejection ratio.

image rejection ratio. In reception using heterodyning in the tuning process, the ratio of (a) the intermediate-frequency (IF) signal level produced by the desired input frequency to (b) that produced by the image frequency. *Note 1:* The image rejection ratio is usually expressed in dB. *Note 2:* When the image rejection ratio is measured, the input signal levels of the desired and image frequencies must be equal for the measurement to be meaningful. *Synonym* image frequency rejection ratio.

imagery. Collectively, the representations of objects reproduced electronically or by optical means on film, electronic display devices, or other media.

IMD. *Abbreviation for* intermodulation distortion.

immediate message. A category of precedence reserved for messages relating to situations that gravely affect the security of national/allied forces or populace and that require immediate delivery to the addressee(s).

IMP. *Abbreviation for* interface message processor.

impedance. The total passive opposition offered to the flow of electric current. *Note 1:* Impedance is determined by the particular combination of resistance, inductive reactance, and capacitive reactance in a given circuit. *Note 2:* Impedance is a function of frequency, except in the case of purely resistive networks.

impedance matching. The connection of an additional impedance to an existing one in order to accomplish a specific effect, such as to balance a circuit or to reduce reflection in a transmission line.

improper character. *Synonym* illegal character.

improved-definition television (IDTV). Television transmitters and receivers that (a) are built to satisfy performance requirements over and above those required by the NTSC standard and (b) remain within the general parameters of NTSC standard emissions. *Note 1:* IDTV improvements may be made at the TV transmitter or the receiver. *Note 2:* Examples of improvements include enhancements in encoding, digital filtering, scan interpolation, interlaced line scanning, and ghost cancellation. *Note 3:* IDTV improvements must allow the TV signal to be transmitted and received in the standard 4:3 aspect ratio. *Synonym* enhanced-quality television.

improvement threshold. *See* FM improvement threshold.

impulse. A short surge of electrical, magnetic, or electromagnetic energy. *Synonym* surge.

impulse excitation. The production of oscillation in a circuit or device by impressing a stimulus (signal) for a period that is extremely short compared to the duration of the oscillation that it produces. *Synonym* shock excitation.

impulse noise. Noise consisting of random occurrences of energy spikes having random amplitude and and spectral content. *Note:* Impulse noise in a data channel can be a definitive cause of data transmission errors.

impulse response. 1. Of a device, the mathematical function that describes the output waveform that results when the input is excited by a unit impulse. 2. The waveform that results at the output of a device when the input is excited by a unit impulse.

IN. *Abbreviation for* intelligent network.

in-band noise power ratio. For multichannel equipment, the ratio of (a) the mean noise power measured in any channel, with all channels loaded with white noise, to (b) the mean noise power measured in the same channel, with all channels but the measured channel loaded with white noise.

in-band signaling. Signaling that uses frequencies or time slots within the bandwidth or data stream occupied by the information channel.

incidental-radiation device. A device that radiates radio frequency energy during the course of its operation although the device is not intentionally designed to generate radio frequency energy.

inclination of an orbit (of an Earth satellite). The angle determined by the plane containing the orbit and the plane of the Earth's equator.

inclined orbit. Any nonequatorial orbit of a satellite. *Note:* Inclined orbits may be circular or elliptical, synchronous or asynchronous, and direct or retrograde.

inclusion. A foreign object present within, for example, an optical fiber or a crystal.

incoherent. In optics, characterized by a degree of coherence significantly less than 0.88.

incoming call identification (ICI). A switching system feature that allows an attendant to identify visually the type of service or trunk group associated with a call directed to the attendant's position.

incorrect block. A block successfully delivered to the intended destination user, but having one or more incorrect bits, additions, or deletions, in the delivered block.

incremental compaction. Data compaction accomplished by specifying only the initial value and all subsequent changes. *Note:* An example of incremental compaction is the storing or transmitting of a line voltage followed only by the deviations from the initial value. Thus, instead of transmitting the values 102, 104, 105, 103, 100, 104 and 106, only the values 102, +2, +1, -2, -3, +4, and +2, or only the values 100, +2, +4, +5, +3, 0, +4, and +6 need be sent, depending on the system used. At a given data rate, transmitting only the initial and incremental values require much less time and space than transmitting the absolute values.

incremental phase modulation (IPM). In spread-spectrum systems, phase modulation in which one binary code sequence is shifted with respect to another, usually to conduct a synchronizing search, *i.e.*, a search to discover if the two sequences are the same, and perhaps thereby enabling two data streams to be synchronized.

indefinite call sign. 1. A call sign that represents a group of facilities, commands, authorities, activities, or units rather than one of these. 2. In radio communications, a call sign that does not identify a station and that is used in the call-up signal or in a message that has the station call sign encrypted in the text.

independent clocks. In communication network timing subsystems, free-running precision clocks used, for synchronization purposes, at the nodes. *Note:* Variable storage buffers, installed to accommodate variations in transmission delay between nodes, are made large enough to accommodate small time (phase) departures among the nodal clocks that control transmission. Traffic may occasionally be interrupted to allow the buffers to be emptied of some or all of their stored data.

independent-sideband (ISB) transmission. Double-sideband transmission in which the information carried by each sideband is different. *Note:* The carrier may be suppressed.

index dip. In an optical fiber, an undesired decrease in the refractive index at the center of the core. *Note:* An index dip is an artifact of certain manufacturing processes. *Synonym* profile dip.

indexing. *See* interaction crosstalk.

index-matching gel. *Synonym* gel (def. #2).

index-matching material. A substance, usually a liquid, cement (adhesive), or gel, which has an index of refraction that closely approximates that of an optical fiber, and is used to reduce Fresnel reflection at the fiber endface. *Note 1:* An index-matching material may be used in conjunction with pairs of mated connectors, with mechanical splices, or at the ends of fibers. *Note 2:* Without the use of an index-matching material, Fresnel reflections will occur at the smooth endfaces of a fiber. These reflections may be as high as –14 dB (*i.e.,* 14 dB below the level of the incident signal). When the reflected signal returns to the transmitting end, it is reflected again and returns to the receiving end at a level that is (28 plus twice the fiber loss) dB below the direct signal. The reflected signal will also be delayed by twice the delay time introduced by the fiber. The reflected signal will have no practical effect on digital systems because of its low level relative to the direct signal; *i.e.*, it will have no practical effect on the detected signal seen at the decision point of the digital optical receiver. It may be noticeable in an analog baseband intensity-modulated video signal.

index of cooperation. 1. In facsimile, the product of the total line length and the number of lines per unit length, divided by π. 2. For rotating devices, the product of the drum diameter and the number of lines per unit length. *Synonyms:* diametral index of cooperation, international index of cooperation.

index of refraction. *Synonym* refractive index.

index profile. *Synonym* refractive index profile.

indirect control. In digital data transmission, the use of a clock rate of 2^n times the modulation rate, where n is an integer greater than one.

indirect wave. A wave, such as a radio wave or sound wave, that arrives at a given point by reflection or scattering from surrounding objects, rather than directly from the source.

individual line. A line that connects a single user to a switching center.

individual reception (in the broadcasting-satellite service). The reception of emissions from a space station in the broadcasting-satellite service by simple domestic installations and in particular those possessing small antennae.

inductive coupling. The transfer of energy from one circuit to another by virtue of the mutual inductance between the circuits. *Note 1:* Inductive coupling may be deliberate and desired (as in an antenna coupler) or may be undesired (as in power line inductive coupling into telephone lines). *Note 2:* Capacitive coupling favors transfer of higher frequency components, whereas inductive coupling favors transfer of lower frequency components.

industrial, scientific, and medical (ISM) applications (of radio frequency energy). Operation of equipment or appliances designed to generate and use locally radio-frequency energy for industrial, scientific, medical, domestic or similar purposes, excluding applications in the field of telecommunications.

industry standard. A voluntary, industry-developed document that establishes requirements for products, practices, or operations.

information. 1. The meaning that a human assigns to data by means of the known conventions used in their representation. 2. In intelligence usage, unprocessed data of every description which may be used in the production of intelligence.

information-bearer channel. 1. A channel capable of transmitting all the information required for communication, such as user data, synchronizing sequences, and control signals. *Note:* The information-bearer channel may operate at a higher data rate than that required for user data alone. 2. A basic communications channel with the necessary bandwidth but without enhanced or value-added services.

information bit. *See* user information bit.

information feedback. The return of received data to the source, usually for the purpose of checking the accuracy of transmission by comparison with the original data.

information field. In data transmission, a field assigned to contain user information. *Note:* The contents of the information field are not interpreted at the link level.

information processing. *Synonym* data processing.

information processing center (IPC). A facility staffed and equipped for processing and distributing information. *Note:* An IPC may be geographically distributed.

information security. The protection of information against unauthorized disclosure, transfer, modification, or destruction, whether accidental or intentional.

information source. *Synonym* source user.

information superhighway. *Synonym* National Information Infrastructure.

information system. 1. A system, whether automated or manual, that comprises people, machines, and/or methods organized to collect, process, transmit, and disseminate data that represent user information. 2. Any telecommunications and/or computer related equipment or interconnected system or subsystems of equipment that is used in the acquisition, storage, manipulation, management, movement, control, display, switching, interchange, transmission, or reception of voice and/or data, and includes software, firmware, and hardware.

information systems security (INFOSEC). The protection of information systems against unauthorized access to or modification of information, whether in storage, processing or transit, and against the denial of service to authorized users or the provision of service to unauthorized users, including those measures necessary to detect, document, and counter such threats.

information transfer. The process of moving messages containing user information from a source to a sink. *Note:* The information transfer rate may or may not be equal to the transmission modulation rate.

information-transfer phase. In an information-transfer transaction, the phase during which user information blocks are transferred from the source user to a destination user.

information-transfer transaction. A coordinated sequence of user and telecommunications system actions that cause information present at a source user to become present at a destination user. *Note:* An information-transfer transaction usually consists of three consecutive phases called the access phase, the information-transfer phase, and the disengagement phase.

INFOSEC. *Acronym for* information systems security.

infrared (IR). The region of the electromagnetic spectrum bounded by the long-wavelength extreme of the visible spectrum (approximately 0.7 μm) and the shortest microwaves (approximately 0.1 μm).

inhibiting signal. A signal that prevents the occurrence of an event. *Note:* An inhibiting signal may be used, for example, to disable an AND gate, thus preventing any signals from passing through it as long as the inhibiting signal is present.

injection fiber. *Synonym* launching fiber.

injection laser diode (ILD). A laser that uses a forward-biased semiconductor junction as the active medium. *Note:* Stimulated emission of coherent light occurs at a p-n junction where electrons and holes are driven into the junction. *Synonyms* diode laser, laser diode, semiconductor laser.

ink vapor recording. Recording in which vaporized ink particles are deposited directly upon the record medium.

input. 1. In a device, process, or channel, a point that accepts data. 2. A state, or a sequence of states, of a point that accepts data. 3. A stimulus, such as a signal or interference, that enters a functional unit, such as a telecommunications system, a computer, or a computer program.

input data. 1. Data being received or to be received by a device or a computer program. 2. Data to be processed.

input-output channel. For a computer, a device that handles the transfer of data between internal memory and peripheral equipment.

input-output controller (IOC). A functional unit that controls one or more input-output channels. *Synonym* I/O controller.

input/output (I/O) device. A device that introduces data into or extracts data from a system.

input protection. For analog input channels, protection against overvoltages that may be applied between any two input connectors or between any input connector and ground.

insertion gain. The gain resulting from the insertion of a device in a transmission line, expressed as the ratio of the signal power delivered to that part of the line following the device to the signal power delivered to that same part before insertion. *Note 1:* If the resulting number is less than unity, an *"insertion loss"* is indicated. *Note 2:* Insertion gain is usually expressed in dB.

insertion loss. 1. The loss resulting from the insertion of a device in a transmission line, expressed as the reciprocal of the ratio of the signal power delivered to that part of the line following the device to the signal power delivered to that same part before insertion. *Note:* Insertion loss is usually expressed in dB. 2. In an optical fiber system, the total optical power loss caused by insertion of an optical component, such as a connector, splice, or coupler.

insertion-loss-vs.-frequency characteristic. Of a system or device, a plot of the amplitude as a function of frequency. *Note:* The insertion-loss-vs.-frequency characteristic may be expressed as absolute gain or loss, or it may be normalized with respect to gain or loss at a specified reference frequency.

inside call. *Synonym* internal call.

inside plant. 1. All the cabling and equipment installed in a telecommunications facility, including the main distribution frame (MDF) and all the equipment extending inward therefrom, such as PABX or central office equipment, MDF protectors, and grounding systems. 2. In radio and radar systems, all communications-electronics (C-E) equipment that is installed in buildings.

inside wire. *See* on-premises wiring.

in-slot signaling. Signaling performed in the associated channel time slot.

inspection lot. A collection of produced units from which a statistically valid sample is to be drawn and inspected to determine conformance with acceptability criteria. *Note:* The inspection lot may differ from a collection of units designated as a lot for other purposes, such as for production, storage, packaging, and shipment.

instrument landing system (ILS). 1. A radio-navigation system which provides aircraft with horizontal and vertical guidance just before and during landing and, at certain fixed points, indicates the distance to the reference point of landing. 2. A system of radio navigation intended to assist aircraft in landing which provides lateral and vertical guidance, which may include indications of distance from the optimum point of landing.

instrument landing system glide path. A system of vertical guidance embodied in the instrument landing system which indicates the vertical deviation of the aircraft from its optimum path of descent.

instrument landing system localizer. A system of horizontal guidance embodied in the instrument landing system which indicates the horizontal deviation of the aircraft from its optimum path of descent along the axis of the runway.

integrated circuit (IC). An electronic circuit that consists of many individual circuit elements, such as transistors, diodes, resistors, capacitors, inductors, and other active and passive semiconductor devices, formed on a single chip of semiconducting material and mounted on a single piece of substrate material. *Synonyms* chip (def.#1), microcircuit.

integrated digital network (IDN). A network that uses both digital transmission and digital switching.

integrated optical circuit (IOC). A circuit, or group of interconnected circuits, consisting of miniature solid-state optical components on semiconductor or dielectric substrates. *Note:* IOC components include light sources, optical filters, photodetectors, and thin-film optical waveguides.

integrated services digital network. *See* ISDN.

integrated station. A terminal device in which a telephone and one or more other devices, such as a video display unit, keyboard, or printer, are integrated and used over a single circuit.

integrated system. A telecommunication system that transfers analog and digital traffic over the same switched network.

integrated voice and data terminal (IVDT). *See* integrated station.

integrating network. A network (circuit) that produces an output waveform that is the time integral of the input waveform. *Note:* Integrating networks are used in signal processing, such as for producing sawtooth waves from square waves.

integrity. *See* data integrity, service integrity, and system integrity.

intelligent network (IN). 1. A network that allows functionality to be distributed flexibly at a variety of nodes on and off the network and allows the architecture to be modified to control the services. 2. In North America, an advanced network concept that is envisioned to offer such things as (a) distributed call-processing capabilities across multiple network modules, (b) real-time authorization code verification, (c) one-number services, and (d) flexible private network services [including (1) reconfiguration by subscriber, (2) traffic analyses, (3) service restrictions, (4) routing control, and (5) data on call histories]. Levels of IN development are identified below:

IN/1 A proposed intelligent network targeted toward services that allow increased customer control and that can be provided by centralized switching vehicles serving a large customer base.

IN/1+ A proposed intelligent network targeted toward services that can be provided by centralized switching vehicles, *e.g.*, access tandems, serving a large customer base.

IN/2 A proposed, advanced intelligent-network concept that extends the distributed IN/1 architecture to accommodate the concept called "*service independence.*" *Note:* Traditionally, service logic has been localized at individual switching systems. The IN/2 architecture provides flexibility in the placement of service logic, requiring the use of advanced techniques to manage the distribution of both network data and service logic across multiple IN/2 modules.

intelligent peripheral (IP). 1. A functional component that may be used most efficiently when accessed locally. 2. An intelligent-network feature that provides specialized telecommunication capabilities required by IN/2 service logic programs.

intelligibility. For voice communications, the capability of being understood. *Note:* Intelligibility does not imply the recognition of a particular voice.

intelligible crosstalk. Crosstalk from which information can be derived.

intensity. The square of the electric field strength of an electromagnetic wave. *Note:* Intensity is proportional to irradiance and may be used in place of the term "*irradiance*" when only relative values are important.

intensity modulation (IM). In optical communications, a form of modulation in which the optical power output of a source is varied in accordance with some characteristic of the modulating signal. *Note:* In intensity modulation, there are no discrete upper and lower sidebands in the usually understood sense of these terms, because present optical sources lack sufficient coherence to produce them. The envelope of the modulated optical signal is an analog of the modulating signal in the sense that the instantaneous power of the envelope is an analog of the characteristic of interest in the modulating signal. Recovery of the modulating signal is by direct detection, not heterodyning.

interaction crosstalk. Crosstalk caused by coupling between carrier and noncarrier circuits. *Note:* If the interaction crosstalk is, in turn, coupled to another carrier circuit, that crosstalk is called "indexing."

interactive data transaction. A unidirectional message, transmitted via a data channel, that requires a response in order for work to proceed logically.

interactive service. In an integrated services digital network (ISDN), a telecommunications service that facilitates a bidirectional exchange of information among users or among users and hosts. *Note:* Interactive services are grouped into conversational services, messaging services, and retrieval services.

interblock gap. On a data recording medium, an area used to indicate the end of a block or physical record. *Note:* Examples of interblock gaps are the gaps between blocks on magnetic tape and disks.

intercept. 1. To stop a telephone call directed to an improper, disconnected, or restricted telephone number, and to redirect that call to an operator or a recording. 2. To gain possession of communications intended for others without their consent, and, ordinarily, without delaying or preventing the transmission. *Note:* An intercept may be an authorized or unauthorized action. 3. The acquisition of a transmitted signal with the intent of delaying or eliminating receipt of that signal by the intended destination user.

interchangeability. A condition which exists when two or more items possess such functional and physical characteristics as to be equivalent in performance and durability, and are capable of being exchanged one for the other without alteration of the items themselves, or of adjoining items, except for adjustment, and without selection for fit and performance.

interchangeable connectors. Connectors that share common installation geometry and have the same transmission performance.

interchange circuit. A circuit that facilitates the exchange of data and signaling information between data terminal equipment (DTE) and data circuit-terminating equipment (DCE). *Note:* An interchange circuit can carry many types of signals and provide many types of service features, such as control signals, timing signals, and common return functions.

intercharacter interval. In asynchronous transmission, the time interval between the end of the stop signal of one character and the beginning of the start signal of the next character. *Note:* The intercharacter interval may be of any duration. The signal sense of the intercharacter interval is always the same as the sense of the stop element, *i.e.*, "1" or "mark."

intercom. 1. A telephone apparatus by means of which personnel can talk to each other within an aircraft, tank, ship, or activity. 2. A dedicated voice service within a specified user environment.

interconnect facility. In a communications network, one or more communications links that (a) are used to provide local area communications service among several locations and (b) collectively form a node in the network. *Note 1:* An interconnect facility may include network control and administrative circuits as well as the primary traffic circuits. *Note 2:* An interconnect facility may use any medium available and may be redundant.

interconnection. 1. The linking together of interoperable systems. 2. The linkage used to join two or more communications units, such as systems, networks, links, nodes, equipment, circuits, and devices.

interexchange carrier (IXC). A communications common carrier that provides telecommunications services between LATAs or between exchanges within the same LATA. *Note:* Interexchange carriers have usually relied on local exchange carriers or competitive access providers for the local origination and termination of their traffic.

interface (I/F). 1. In a system, a shared boundary, *i.e.,* the boundary between two subsystems or two devices. 2. A shared boundary between two functional units, defined by specific attributes, such as functional characteristics, common physical interconnection characteristics, and signal characteristics. 3. A point of communication between two or more processes, persons, or other physical entities. 4. A point of interconnection between user terminal equipment and commercial communications facilities. 5. To interconnect two or more entities at a common point or shared boundary.

interface functionality. In telephony, the characteristic of interfaces that allows them to support transmission, switching, and signaling functions identical to those used in the enhanced services provided by the carrier. *Note:* As part of its comparably efficient interconnection (CEI) offering, the carrier must make available standardized hardware and software interfaces that are able to support transmission, switching, and signaling functions identical to those used in the enhanced services provided by the carrier.

interface message processor (IMP). A processor-controlled switch used in packet-switched networks to route packets to their proper destination.

interface payload. In integrated services digital networks (ISDN), the part of the bit stream through a framed interface used for telecommunications services and signaling.

interface point. *Synonym* **point of interface.**

interface standard. A standard that describes one or more functional characteristics (such as code conversion, line assignments, or protocol compliance) or physical characteristics (such as electrical, mechanical, or optical characteristics) necessary to allow the exchange of information between two or more (usually different) systems or equipment. *Note 1:* An interface standard may include operational characteristics and acceptable levels of performance. *Note 2:* In the military community, interface standards permit command and control functions to be performed using communication and computer systems.

interference. 1. In general, extraneous energy, from natural or man-made sources, that impedes the reception of desired signals. 2. A coherent emission having a relatively narrow spectral content, *e.g.,* a radio emission from another transmitter at approximately the same frequency, or having a harmonic frequency approximately the same as, another emission of interest to a given recipient, and which impedes reception of the desired signal by the intended recipient. *Note:* In the context of this definition, interference is distinguished from noise in that the latter is an incoherent emission from a natural source *(e.g.,* lightning) or a man-made source, of a character unlike that of the desired signal (*e.g.,* commutator noise from rotating machinery) and which usually has a broad spectral content. 3. The effect of unwanted energy due to one or a combination of emissions, radiation, or inductions upon reception in a radiocommunication system, manifested by any performance degradation, misinterpretation, or loss of information which could be extracted in the absence of such unwanted energy. 4. The interaction of two or more coherent or partially coherent waves, which interaction produces a resultant wave that differs from the original waves in phase, amplitude, or both. *Note:* Interference may be constructive or destructive, *i.e.,* it may result in increased amplitude or decreased amplitude, respectively. Two waves equal in frequency and amplitude, and out of phase by 180°, will completely cancel one another. In phase, they create a resultant wave having twice the amplitude of either interfering beam.

interference emission. Emission that results in an electrical signal's being propagated into, and interfering with the proper operation of, electronic or electrical equipment. *Note:* The frequency range of interference emissions may include the entire electromagnetic spectrum.

interference filter. An optical filter that reflects one or more spectral bands or lines and transmits others, while maintaining a nearly zero coefficient of absorption for all wavelengths of interest. *Note 1:* An interference filter may be high-pass, low-pass, bandpass, or band-rejection. *Note 2:* An interference filter consists of multiple thin layers of dielectric material having different refractive indices. There also may be metallic layers. Interference filters are wavelength-selective by virtue of the interference effects that take place between the incident and reflected waves at the thin-film boundaries.

interferometer. An instrument that uses the principle of interference of electromagnetic waves for purposes of measurement. *Note:* Interferometers may be used to measure a variety of physical variables, such as displacement (distance), temperature, pressure, and strain.

interferometry. The branch of science devoted to the study and measurement of the interaction of waves, such as electromagnetic waves and acoustic waves. *Note 1:* The interaction of the waves can produce various spatial-, time-, and frequency-domain energy distribution patterns. [After 2196] *Note 2:* Interferometric techniques are used to measure refractive index profiles, *e.g.,* those of the preforms from which optical fibers are drawn, and to sense and measure physical variables, such as displacement (distance), temperature, pressure, and magnetic fields.

interframe time fill. In digital data transmission, a sequence of bits transmitted between consecutive frames. *Note:* Interframe time fill does not include bits stuffed within a frame.

interlaced scanning. In raster-scanned video displays, a scanning technique in which all odd-numbered scanning lines are first traced in succession, followed by the tracing of the even-numbered scanning lines in succession, each of which is traced between a pair of odd-numbered scanning lines. *Note 1:* The pattern created by tracing the odd-numbered scanning lines is called the *"odd field"*, and the pattern created by tracing the even-numbered scanning lines is called the *"even field"*. Each field contains half the information content, *i.e.,* pixels, of the complete video frame. *Note 2:* Image flicker is less apparent in an interlaced display than in a noninterlaced display, because the rate at which successive fields occur in an interlaced display is twice that at which successive frames would occur in a noninterlaced display containing the same number of scanning lines and having the same frame refresh rate. *Synonym* interlacing.

interlacing. *Synonym* interlaced scanning.

inter-LATA. 1. Between local access and transport areas (LATAs). 2. Services, revenues, and functions associated with telecommunications that originate in one LATA and that terminate in another one or that terminate outside of that LATA.

interleaving. The transmission of pulses from two or more digital sources in time-division sequence over a single path.

intermediate distribution frame (IDF). In a central office or customer premises, a frame that (a) cross-connects the user cable media to individual user line circuits and (b) may serve as a distribution point for multipair cables from the main distribution frame (MDF) or combined distribution frame (CDF) to individual cables connected to equipment in areas remote from these frames.

intermediate element. In a network, a line-unit-line termination (LULT) or a line-unit-network termination (LUNT).

intermediate field. *Synonym* intermediate-field region.

intermediate-field region. For an antenna, the transition region–lying between the near-field region and the far-field region–in which the field strength of an electromagnetic wave is dependent upon the inverse distance, inverse square of the distance, and the inverse cube of the distance from the antenna. *Note:* For an antenna that is small compared to the wavelength in question, the intermediate-field region is considered to exist at all distances between 0.1

wavelength and 1.0 wavelength from the antenna. *Synonyms* intermediate field, intermediate zone, transition zone.

intermediate frequency (IF). A frequency to which a carrier frequency is shifted as an intermediate step in transmission or reception.

intermediate language. In computer programming, a target language into which all or part of a single statement or a source program—in a source language—is translated before it is further translated or interpreted. *Note*: For a subsequent translation, an intermediate language may serve as a source language.

intermediate-level language. In computer, communications, and data processing systems, a programming language that (a) is less machine-oriented than a machine language, (b) is not so machine-independent as a common language, such as Ada®, COBOL, or Fortran, (c) contains macros that are less powerful than common-language macros, and (d) usually is the object language of a root compiler. Examples of intermediate-level languages include assembly languages, such as PL/I.

intermediate system. A system that provides an Open Systems Interconnection—Reference Model (OSI—RM) Network Layer relay function in which data received from one corresponding network entity are forwarded to another corresponding network entity.

intermediate zone. *Synonym* intermediate-field region.

intermodal delay distortion. *Synonym* multimode distortion.

intermodal dispersion. *Incorrect synonym for* multimode distortion.

intermodal distortion. *Synonym* multimode distortion.

intermodulation (IM). The production, in a nonlinear element of a system, of frequencies corresponding to the sum and difference frequencies of the fundamentals and harmonics thereof that are transmitted through the element.

intermodulation distortion. Nonlinear distortion characterized by the appearance, in the output of a device, of frequencies that are linear combinations of the fundamental frequencies and all harmonics present in the input signals. *Note:* Harmonic components themselves are not usually considered to characterize intermodulation distortion. When the harmonics are included as part of the distortion, a statement to that effect should be made.

intermodulation noise. In a transmission path or device, noise, generated during modulation and demodulation, that results from nonlinear characteristics in the path or device.

intermodulation product. In the output of a nonlinear system, a frequency produced by intermodulation of harmonics of the frequencies present in the input signal.

internal bias. In a start-stop teletypewriter receiving mechanism, bias generated locally by the mechanism, and which has the same effect on the operating margin as bias external to the receiver, *i.e.*, applied bias. *Note:* Internal bias may be a marking bias or a spacing bias.

internal call. A call placed within a private branch exchange (PBX) or local switchboard, *i.e.*, not through a central office in a public switched network. *Synonym* inside call.

internal memory. In a computer, all of the storage spaces that are accessible by a processor without the use of the computer input-output channels. *Note:* Internal memory usually includes several types of storage, such as main storage, cache memory, and special registers, all of which can be directly accessed by the processor. *Synonym* internal storage.

internal photoelectric effect. A photoconductive or photovoltaic effect.

internal storage. *Synonym* internal memory.

International Atomic Time (TAI). The time scale established by the International Time Bureau (BIH) on the basis of atomic clock data supplied by cooperating institutions. *Note:* The abbreviations "TAI" and "BIH" are a result of literal translation from the official international names that are written in French.

international index of cooperation. *Synonym* index of cooperation.

International Frequency Registration Board (IFRB). *See* Radio Regulations Board.

International Organization for Standardization. *See* ISO.

International Radio Consultative Committee. *See* CCIR, ITU-R.

International System of Units (SI). The modern form of the metric system, which has been adopted by the United States and most other nations. *Note:* The SI is constructed from seven base units for independent physical quantities. *Tables showing these values are included on the next pages and are current as of Fall 1995.*

SI prefixes. The common metric prefixes are:

Multiplication Factor		*Prefix*	*Nam*	*Prefix Symbol*
1 000 000 000 000	=	10^{12}	tera	T
1 000 000 000	=	10^{9}	giga	G
1 000 000	=	10^{6}	mega	M
1 000	=	10^{3}	kilo	k
100	=	10^{2}	hecto	h
10	=	10^{1}	deka	da
0.1	=	10^{-1}	deci	d
0.01	=	10^{-2}	centi	c
0.001	=	10^{-3}	milli	m
0.000 001	=	10^{-6}	micro	µ
0.000 000 001	=	10^{-9}	nano	n
0.000 000 000 001	=	10^{-12}	pico	p

SI Base Units

Quantity	*Unit Name*	*Unit Symbol*
length	meter	m
mass	kilogram	kg
time	second	s
electric current	ampere	A
thermodynamic temperature	kelvin	K
amount of substance	mole	mol
luminous intensity	candela	cd

SI derived units. Derived units are formed by combining base units and other derived units according to the algebraic relations linking the corresponding quantities. The symbols for derived units are obtained by means of the mathematical signs for multiplication, division, and use of exponents. For example, the SI unit for velocity is the *meter per second* (m/s or m•s^{-1}), and that for angular velocity is the *radian per second* (rad/s or rad•s^{-1}).

International Telecommunication Union (ITU). A civil international organization established to promote standardized telecommunications on a worldwide basis. *Note:* The ITU-R and ITU-T are committees under the ITU. The ITU headquarters is located in Geneva, Switzerland. While older than the United Nations, it is recognized by the U.N. as the specialized agency for telecommunications.

International Telegraph Alphabet Number 5 (ITA-5). An alphabet in which (a) 128 unique 7-bit strings are used to encode upper- and lower-case letters, 10 decimal numerals, special signs and symbols, diacritical marks, data delimiters, and transmission control characters, (b) 12 of the 7-bit strings are not assigned to any letter, numeral, or control character, and (c) the unassigned bit strings are open for use in a given country that may have unique requirements, such as monetary symbols; diacritical marks, such as the tilde, umlaut, circumflex, and dieresis, and (d) a two-condition 8-bit pattern may be used that consists of seven information bits and a parity check bit. *Note:* ITA-5 is used for effecting information interchange. It is a result of a joint agreement between the International Telegraph and Telephone Consultative Committee (CCITT), now ITU-T, of the International Telecommunication Union (ITU) and the International Organization for Standardization (ISO). It is published as CCITT Recommendation V.3 and as ISO 646. It has also been adopted by NATO for military use. The United States adaptation of ITA-5 is ASCII (American Standard Code for Information Interchange) published by the American National Standards Institute (ANSI).

International Telegraph and Telephone Consultative Committee. *See* CCITT, ITU-T.

International Time Bureau (BIH). *See* International Atomic Time.

internet. An interconnection of networks.

[The] Internet. A worldwide interconnection of individual networks operated by government, industry, academia, and private parties. *Note:* The Internet originally served to interconnect laboratories engaged in government research, and has now been expanded to serve millions of users and a multitude of purposes.

Internet protocol (IP). A DOD standard protocol designed for use in interconnected systems of packet-switched computer communication networks. *Note:* The internet protocol provides for transmitting blocks of data called *datagrams* from sources to destinations, where sources and destinations are hosts identified by fixed-length addresses. The internet protocol also provides for fragmentation and reassembly of long datagrams, if necessary, for transmission through small-packet networks.

Internet protocol (IP) spoofing. 1. The creation of IP packets with counterfeit (spoofed) IP source addresses. 2. A method of attack used by network intruders to defeat network security measures such as authentication based on IP addresses. *Note 1:* An attack using IP spoofing may lead to unauthorized user access, and possibly root access, on the targeted system. *Note 2:* A packet-filtering-router firewall may not provide adequate protection against IP spoofing attacks. It is possible to route packets through this type of firewall if the router is not configured to filter incoming packets having source addresses on the local domain. *Note 3:* IP spoofing is possible even if no reply packets can reach the attacker. *Note 4:* A method for preventing IP spoofing problems is to install a filtering router that does not allow incoming packets to have a source address different from the local domain. In addition, outgoing packets should not be allowed to contain a source address different from the local domain, in order to prevent an IP spoofing attack from originating from the local network.

internetwork connection. *See* gateway.

internetworking. The process of interconnecting two or more individual networks to facilitate communications among their respective nodes. *Note:* The interconnected networks may be different types. Each network is distinct, with its own addresses, internal protocols, access methods, and administration.

interoffice trunk. A single direct transmission channel, *e.g.,* voice-frequency circuit, between central offices.

interoperability. 1. The ability of systems, units, or forces to provide services to and accept services from other systems, units or forces and to use the services so exchanged to enable them to operate effectively together. 2. The condition achieved among communications-electronics systems or items of communications-electronics equipment when information or services can be exchanged directly and satisfactorily between them and/or their users. The degree of interoperability should be defined when referring to specific cases.

interoperability standard. A document that establishes engineering and technical requirements that are necessary to be employed in the design of systems, units, or forces and to use the services so exchanged to enable them to operate effectively together.

interoperation. The use of interoperable systems, units, or forces.

interposition trunk. 1. A single direct transmission channel, *e.g.,* voice-frequency circuit, between two positions of a large switchboard to facilitate the interconnection of other circuits appearing at the respective switchboard positions. 2. Within a technical control facility, a single direct transmission circuit, between positions in a testboard or patch bay, which circuit facilitates testing or patching between the respective positions.

interpret. To translate and to execute each source language statement of a computer program before translating and executing the next statement.

interrogation. 1. The transmission of a signal or combination of signals intended to trigger a response. 2. The process whereby a station or device requests another station or device to identify itself or to give its status.

interrupt. A suspension of a process, such as the execution of a computer program, caused by an event external to that process, and performed in such a way that the process can be resumed. *Synonym* **interruption.**

interrupted continuous wave (ICW). Modulation in which there is on-off keying of a continuous wave.

interrupted isochronous transmission. *Synonym* isochronous burst transmission.

interruption. *Synonym* interrupt.

inter-satellite service. A radiocommunication service providing links between artificial Earth satellites.

interswitch trunk. A single direct transmission channel, *e.g.,* voice-frequency circuit, between switching nodes.

intersymbol interference. 1. In a digital transmission system, distortion of the received signal, which distortion is manifested in the temporal spreading and consequent overlap of individual pulses to the degree that the receiver cannot reliably distinguish between changes of state, *i.e.*, between individual signal elements. *Note 1:* At a certain threshold, intersymbol interference will compromise the integrity of the received data. *Note 2:* Intersymbol interference attributable to the statistical nature of quantum mechanisms sets the fundamental limit to receiver sensitivity. *Note 3:* Intersymbol interference may be measured by eye patterns. 2. Extraneous energy from the signal in one or more keying intervals that interferes with the reception of the signal in another keying interval. 3. The disturbance caused by extraneous energy from the signal in one or more keying intervals that interferes with the reception of the signal in another keying interval.

intertoll trunk. A single direct transmission channel, *e.g.,* voice-frequency circuit, between two toll offices.

interworking functions. Mechanisms that mask differences in physical, link, and network technologies by converting (or mapping) states and protocols into consistent network and user services.

intra-LATA. Within the boundaries of a local access and transport area (LATA).

intramodal distortion. In an optical fiber, distortion caused by dispersion, such as material or profile dispersion, of a given propagating mode.

intraoffice trunk. A single direct transmission channel, *e.g.,* voice-frequency circuit, within a given switching center.

intrinsic joint loss. Of nonidentical optical fibers joined by a splice or a mated pair of connectors, the power loss attributable to manufacturing variations, in such parameters as physical dimensions, differences in refractive index (including profile parameter), numerical aperture, and mode field diameter.

intrinsic noise. In a transmission path or device, that noise inherent to the path or device and not contingent upon modulation.

inverse multiplexer. A functional unit capable of accessing and combining two or more low-speed circuits into a virtual broadband circuit, up to and including an aggregate equal to a T1 rate.

inverse-square law. The physical law stating that irradiance, *i.e.*, the power per unit area in the direction of propagation, of a spherical wavefront varies inversely as the square of the distance from the source, assuming there are no losses caused by absorption or scattering. *Note:* For example, the power radiated from a point source, *e.g.*, an omnidirectional isotropic antenna, or from any source at very large distances from the source compared to the size of the source, must spread itself over larger and larger spherical surfaces as the distance from the source increases. Diffuse and incoherent radiation are similarly affected.

inverted position. In frequency-division multiplexing, a position of a translated channel in which an increasing signal frequency in the untranslated channel causes a decreasing signal frequency in the translated channel.

inverter. 1. In electrical engineering, a device for converting direct current into alternating current. 2. In computers, a device or circuit that inverts the polarity of a signal or pulse. *Deprecated synonym* negation circuit.

Inward Wide-Area Telephone Service (INWATS). *See* eight-hundred (800) service.

INWATS. *Acronym for* Inward Wide-Area Telephone Service. *See* eight-hundred (800) service.

I/O. *Abbreviation for* input/output.

IOC. *Abbreviation for* input-output controller, integrated optical circuit.

I/O controller. *Synonym* input-output controller.

ionosphere. That part of the atmosphere, extending from about 70 to 500 kilometers, in which ions and free electrons exist in sufficient quantities to reflect and/or refract electromagnetic waves.

ionosphere sounder. A device that transmits signals for the purpose of determining ionospheric conditions.

ionospheric absorption. Absorption occurring as a result of interaction between an electromagnetic wave and free electrons in the ionosphere.

ionospheric disturbance. An increase in the ionization of the ionosphere, caused by solar activity, which results in greatly increased radio wave absorption.

ionospheric forward scatter (IFS). *Synonym* ionospheric scatter.

ionospheric reflection. Of electromagnetic waves propagating in the ionosphere, a redirection, *i.e.,* bending—by a complex processing involving reflection and refraction—of the waves back toward the Earth. *Note:* The amount of bending depends on the extent of penetration (which is a function of frequency), the angle of incidence, polarization of the wave, and ionospheric conditions, such as the ionization density.

ionospheric scatter. The propagation of radio waves by scattering as a result of irregularities or discontinuities in the ionization of the ionosphere. *Synonym* forward propagation ionospheric scatter.

ionospheric sounding. A technique that provides real-time data on high-frequency ionospheric-dependent radio propagation, using a basic system consisting of a synchronized transmitter and receiver. *Note:* The time delay between transmission and reception is translated into effective ionospheric layer altitude. Vertical incident sounding uses a collocated transmitter and receiver and involves directing a range of frequencies vertically to the ionosphere and measuring the values of the reflected returned signals to determine the effective ionosphere layer altitude. This technique is also used to determine the critical frequency. Oblique sounders use a transmitter at one end of a given propagation path, and a synchronized receiver, usually with an oscilloscope-type display (ionogram), at the other end. The transmitter emits a stepped- or swept-frequency signal which is displayed or measured at the receiver. The measurement converts time delay to effective altitude of the ionospheric layer. The ionogram display shows the effective altitude of the ionospheric layer as a function of frequency.

ionospheric turbulence. Ongoing disturbances of the ionosphere that scatter incident electromagnetic waves. *Note:* Ionospheric turbulence results in irregularities in the composition of the ionosphere that change with time. This causes changes in reflection properties. These, in turn, cause changes in skip distance, fading, local intensification, and distortion of the incident waves.

IP. *Abbreviation for* intelligent peripheral, Internet protocol.

IPC. *Abbreviation for* information processing center.

IPX. *Abbreviation for* Internetwork Packet Exchange. A proprietary LAN protocol.

IR. *Abbreviation for* infrared.

irradiance. Radiant power incident per unit area upon a surface. *Note:* Irradiance is usually expressed in watts per square meter, but may also be expressed in joules per square meter. *Deprecated synonym* power density.

irradiation. The product of irradiance and time, *i.e.,* the time integral of irradiance. *Note:* For example, an irradiation of 100 J/m^2 (joules per square meter) is obtained when an irradiance of 25 W/m^2 (watts per square meter) is continuously incident for 4 seconds.

ISB. *Abbreviation for* independent sideband. *See* independent-sideband transmission.

ISDN. *Abbreviation for* integrated services digital network. An integrated digital network in which the same time-division switches and digital transmission paths are used to establish connections for different services. *Note 1:* ISDN services include telephone, data, electronic mail, and facsimile. *Note 2:* The method used to accomplish a connection is often specified: for example, switched connection, nonswitched connection, exchange connection, ISDN connection.

ISM. *Abbreviation for* industrial, scientific, and medical applications (of radio frequency energy).

ISO. *Abbreviation for* International Organization for Standardization. An international organization that (a) consists of member bodies that are the national standards bodies of most of the countries of the world, (b) is responsible for the development and publication of international standards in various technical fields, after developing a suitable consensus, (c) is affiliated with the United Nations, and (d) has its headquarters at 1, rue de Varembé, Geneva, Switzerland. *Note:*

Member bodies of ISO include, among others, the American National Standards Institute (ANSI), the Association Française de Normalisation (AFNOR), the British Standards Institution (BSI), and the Deutsche Institut für Normung (DIN).

isochrone. A line on a map or chart joining points associated with a constant time difference from the transmitter to receiver of electromagnetic waves, such as radio waves, at all points along the line.

isochronous. 1. Of a periodic signal, pertaining to transmission in which the time interval separating any two corresponding transitions is equal to the unit interval or to a multiple of the unit interval. 2. Pertaining to data transmission in which corresponding significant instants of two or more sequential signals have a constant phase relationship. *Note:* "Isochronous" and "anisochronous" are characteristics, while "synchronous" and "asynchronous" are relationships.

isochronous burst transmission. In a data network where the information-bearer channel rate is higher than the input data signaling rate, transmission performed by interrupting, at controlled intervals, the data stream being transmitted. *Note 1:* Isochronous burst transmission enables communication between data terminal equipment (DTE) and data networks that operate at dissimilar data signaling rates, such as when the information-bearer channel rate is higher than the DTE output data signaling rate. *Note 2:* The binary digits are transferred at the information-bearer channel rate. The transfer is interrupted at intervals in order to produce the required average data signaling rate. *Note 3:* The interruption is always for an integral number of unit intervals. *Note 4:* Isochronous burst transmission has particular application where envelopes are being transferred between data circuit-terminating equipment (DCE) and only the bytes contained within the envelopes are being transferred between the DCE and the DTE. *Synonyms* burst isochronous *(deprecated)*, interrupted isochronous transmission.

isochronous demodulation. Demodulation in which the time interval separating any two significant instants is equal to the unit interval or a multiple of the unit interval.

isochronous distortion. The difference between the measured modulation rate and the theoretical modulation rate in a digital system.

isochronous modulation. Modulation in which the time interval separating any two significant instants is equal to the unit interval or a multiple of the unit interval.

isochronous signal. A signal in which the time interval separating any two significant instants is equal to the unit interval or a multiple of the unit interval. *Note 1:* Variations in the time intervals are constrained within specified limits. *Note 2:* "Isochronous" is a characteristic, while "synchronous" indicates a relationship.

isolator. *See* optical isolator.

isotropic. 1. Pertaining to a material with properties, such as density, electrical conductivity, electric permittivity, magnetic permeability, or refractive index that do not vary with distance or direction. 2. Pertaining to a material with magnetic, electrical, or electromagnetic properties that do not vary with the direction of static or propagating magnetic, electrical, or electromagnetic fields within the material.

isotropic antenna. A hypothetical antenna that radiates or receives equally in all directions. *Note:* Isotropic antennas do not exist physically but represent convenient reference antennas for expressing directional properties of physical antennas.

isotropic gain. *Synonym* absolute gain (def. #1).

iterative impedance. In electrical circuits, for a four-terminal network, the impedance that, if connected across one pair of terminals, will match the impedance across the other pair of terminals. *Note:* The iterative impedance of a uniform line is the same as its characteristic impedance.

ITU. *Abbreviation for* International Telecommunication Union.

ITU-R. The Radiocommunications Sector of the ITU; responsible for studying technical issues related to radiocommunications, and having some regulatory powers. *Note:* A predecessor organization was the CCIR.

ITU-T. *Abbreviation for* International Telecommunication Union—Telecommunication Standardization Bureau. The Telecommunications Standardization Sector of the International Telecommunication Union (ITU). *Note 1:* ITU-T is responsible for studying technical, operating, and tariff Questions and issuing Recommendations on them, with the goal of standardizing telecommunications worldwide. *Note 2:* The ITU-T combines the standards-setting activities of the predecessor organizations formerly called the International Telegraph and Telephone Consultative Committee (CCITT) and the International Radio Consultative Committee (CCIR).

IVDT. *Abbreviation for* integrated voice data terminal. *See* integrated station.

IXC. *Abbreviation for* interexchange carrier.

jabber. In local area networks, transmission by a data station beyond the time interval allowed by the protocol.

jacket. *Synonym* sheath.

jamming margin. The level of interference (jamming) that a system is able to accept and still maintain a specified level of performance, such as maintain a specified bit-error ratio even though the signal-to-noise ratio is decreasing.

jamming to signal ratio (J/S). The ratio, usually expressed in dB, of the power of a jamming signal to that of a desired signal at a given point such as the antenna terminals of a receiver.

jam signal. A signal that carries a bit pattern sent by a data station to inform the other stations that they must not transmit. *Note 1:* In carrier-sense multiple access with collision detection (CSMA/CD) networks, the jam signal indicates that a collision has occurred. *Note 2:* In carrier-sense multiple access with collision avoidance (CSMA/CA) networks, the jam signal indicates that the sending station intends to transmit. *Note 3: "Jam signal"* should not be confused with *"electronic jamming."*

jerkiness. In a video display, the perception, by human vision faculties, of originally continuous motion as a sequence of distinct "snapshots." *Note 1:* The perception of continuous motion by human vision faculties is a manifestation of complex functions, *i.e.,* characteristics, of the eyes and brain. When presented with a sequence of fixed, *i.e.,* still, images of sufficient continuity and at a sufficiently frequent update rate, the brain interpolates intermediate images, and the observer subjectively appears to see continuous motion that in reality does not exist. *Note 2:* For example, the update rate of NTSC television displays is 30 frames (60 fields) per second.

jitter. Abrupt and unwanted variations of one or more signal characteristics, such as the interval between successive pulses, the amplitude of successive cycles, or the frequency or phase of successive cycles. *Note 1:* Jitter must be specified in qualitative terms (*e.g.,* amplitude, phase, pulse width or pulse position) and in quantitative terms (*e.g.,* average, RMS, or peak-to-peak). *Note 2:* The low-frequency cutoff for jitter is usually specified at 1 Hz. *Contrast with* drift, wander.

job. In computing, a unit of work that is defined by a user and that is to be accomplished by a computer. *Note:* A job is identified by a label and usually includes a set of computer programs, files, and control statements to the computer operating system.

job-recovery control file. *Synonym* backup file.

Johnson noise. *Synonym* thermal noise.

joint. For optical fibers, a splice or connector.

joint multichannel trunking and switching system. That composite multichannel trunking and switching system formed from assets of the Services, the Defense Communications System, other available systems, and/

or assets controlled by the Joint Chiefs of Staff to provide an operationally responsive, survivable communication system, preferably in a mobile/ transportable/recoverable configuration, for the joint force commander in an area of operations.

Joint Spectrum Center (JSC). A DOD center that consolidates the former Electromagnetic Compatibility Analysis Center (ECAC) and functions of the military Department Spectrum centers.

Joint Tactical Information Distribution System (JTIDS). An advanced information distribution system that provides secure integrated communication, navigation, and identification (ICNI) capability for application to military tactical operations.

Joint Telecommunications Resources Board (JTRB). The body required to be established by Section 2(b) (3) of Executive Order No. 12472 to assist the Director of the Office of Science and Technology Policy in the exercise of assigned nonwartime emergency telecommunications functions.

journal. 1. A chronological record of data processing operations that may be used to reconstruct a previous or an updated version of a file. *Synonym* log. 2. In database management systems, the record of all stored data items that have values changed as a result of processing and manipulation of the data.

joy stick. In computer graphics, a lever (with at least two degrees of freedom) that is used as an input unit, normally as a locator.

JSC. *Abbreviation for* Joint Spectrum Center.

JTIDS. *Acronym for* Joint Tactical Information Distribution System.

JTRB. *Abbreviation for* Joint Telecommunications Resources Board.

Julian date. 1. The sequential day count reckoned consecutively beginning January 1, 4713 B.C. *Note:* The Julian date on January 1, 1990, was 2,446,892. 2. The sequential day count of the days of a year, reckoned consecutively from the first day of January. *Note:* In modern times, the definition of Julian date has been corrupted to use the first day of the year as the point of reference. To avoid ambiguity with the traditional meaning, *"day of year"* rather than *"Julian date"* should be used for this purpose.

jumper. *Synonym* cross-connection.

junction point. *Synonym* node (def. #1).

justification. *See* bit stuffing, de-stuffing, justify.

justify. 1. To shift the contents of a register or a field so that the significant character at the specified end of the data is at a particular position. 2. To align text horizontally or vertically so that the first and last characters of every line, or the first and last line of the text, are aligned with their corresponding margins. *Note 1:* In English, text may be justified left, right, or both. Left justification is the most common. *Note 2:* The last line of a paragraph is usually only left justified. 3. To align data on a designated character position.

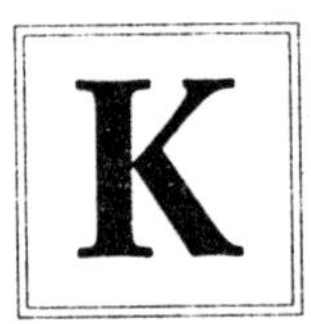

k. *Abbreviation for* kilo (SI prefix for 10^3). *See* metric system.

K. 1. *Abbreviation for* kelvin(s). *See* thermodynamic temperature. 2. When referring to data storage capacity, 2^{10}, or 1024 in decimal notation; however this usage of an upper case K is deprecated.

Kalman filter. A computational algorithm that processes measurements to deduce an optimum estimate of the past, present, or future state of a linear system by using a time sequence of measurements of the system behavior, plus a statistical model that characterizes the system and measurement errors, plus initial condition information.

KDC. *Abbreviation for* key distribution center.

kelvin (K). A unit of thermodynamic temperature, taken as one of the base units of the International System of Units (SI). The kelvin is defined by setting the thermodynamic temperature of the triple point of water at 273.16 K. *Note 1:* The kelvin was formerly called *"degree Kelvin."* The term *"degree Kelvin"* is now obsolete. No degree symbol is written with K, the symbol for kelvin(s). *Note 2:* In measuring temperature intervals, the degree Celsius is equal to the kelvin. The Celsius temperature scale is defined by setting 0 °C equal to 273.16 K.

kelvin temperature scale. *See* thermodynamic temperature.

Kendall effect. A spurious pattern or other distortion in a facsimile record copy caused by unwanted modulation products arising from the transmission of a carrier signal, and appearing in the form of a rectified baseband that interferes with the lower sideband of the carrier. *Note:* The Kendall effect occurs principally when the single-sideband width is greater than half of the facsimile carrier frequency.

Kennelly-Heaviside layer. *Synonym* E region.

kernel. A module of a program that forms a logical entity or performs a unit function. *Note:* The most vulnerable portion of code in a secure operating system is a special case of a kernel.

Kerr electro-optic effect. The creation of birefringence in a liquid that is not otherwise birefringent, by subjecting the liquid to an electric field. *Note 1:* The degree of birefringence, which is manifested as a difference in refractive indices for light of orthogonal linear polarizations, one of which is parallel to the induced optical axis, is directly proportional to the square of the applied electric field strength. *Note 2:* In the general case, the birefringence produced by the applied electric field can be used in conjunction with polarizers to modulate light. Devices that use this principle are called *Kerr cells*.

key. Information (usually a sequence of random or pseudorandom binary digits) used initially to set up and periodically change the operations performed in crypto-equipment for the purpose of encrypting or decrypting electronic signals, for determining electronic counter-countermeasures patterns (*e.g.*, frequency hopping or spread spectrum), or for producing other key. *Note:* "*Key*" has replaced the terms "*variable,*" "*key(ing) variable,*" and "*cryptovariable.*"

keyboard. An input device used to enter data by manual depression of keys, which causes the generation of the selected code element.

keyboard punch. *Synonym* keypunch.

key distribution center (KDC). A COMSEC facility that generates and distributes key in electrical form.

keying. The generating of signals by the interruption or modulation of a steady signal or carrier. *See* chroma keying.

keying variable. *Deprecated synonym for* key.

key management. [The] Process by which key is generated, stored, protected, transferred, loaded, used, and destroyed.

key pulsing. A system of sending telephone calling signals in which the digits are transmitted by operation of a pushbutton key set. *Note:* The type of key pulsing commonly used by users and PBX operators is dual-tone multifrequency signaling. Each pushbutton causes generation of a unique pair of tones. In military systems, pushbuttons are also provided for additional signals, such as precedence. *Synonym* pulsing.

keypunch. A keyboard-actuated punch that punches holes in a data medium. *Synonym* **keyboard punch.**

key set. A multiline or multifunction user terminal device.

key stream. [A] sequence of symbols (or their electrical or mechanical equivalents) produced in a machine or auto-manual cryptosystem to combine with plain text to produce cipher text, controi transmission security processes, or produce key.

key telephone system (KTS). In a local environment, terminals and equipment that provide immediate access from all terminals to a variety of telephone services without attendant assistance. *Note:* A KTS may interface with the public switched telephone network.

key variable. *Deprecated synonym for* key.

k-factor. 1. In tropospheric radio propagation, the ratio of the effective Earth radius to the actual Earth radius. *Note:* The k-factor is approximately 4/3. 2. In ionospheric radio propagation, a correction factor that (a) is applied in calculations related to curved layers, and (b) is a function of distance and the real height of ionospheric reflection.

kHz. *Abbreviation for* kilohertz.

kilohertz (kHz). A unit of frequency denoting one thousand (10^3) Hz.

kilometer. A unit of distance corresponding to 1000m.

km. *Abbreviation for* kilometer.

knife-edge effect. In electromagnetic wave propagation, a redirection by diffraction of a portion of the incident radiation that strikes a well-defined obstacle such as a mountain range or the edge of a building. *Note:* The knife-edge effect is explained by Huygens' principle, which states that a well-defined obstruction to an electromagnetic wave acts as a secondary source, and creates a new wavefront. This new wavefront propagates into the geometric shadow area of the obstacle.

k*T*. *See* noise-power density.

KTS. *Abbreviation for* key telephone system.

K-type patch bay. A patching facility designed for patching and monitoring of balanced digital data circuits that support data rates up to 1 Mb/s.

label. 1. An identifier within or attached to a set of data elements. 2. One or more characters that (a) are within or attached to a set of data elements and (b) represent information about the set, including its identification. 3. In communications, information within a message that is used to identify specific system parameters, such as the particular circuit with which the message is associated. *Note:* Messages that do not relate to call control should not contain a label. 4. In programming languages, an identifier that names a statement. 5. An identifier that indicates the sensitivity of the attached information. 6. For classified information, an identifier that indicates (a) the security level of the attached information or (b) the specific category in which the attached information belongs.

labeled channel. In integrated services digital networks, (ISDN), a time-ordered set of all block payloads that have labels containing the same information, *i.e.*, containing the same identifiers.

labeled interface structure. In integrated services digital networks (ISDN), an interface structure that provides telecommunications services and signaling by means of labeled channels.

labeled multiplexing. In integrated services digital networks (ISDN), multiplexing by concatenation of the blocks of the channels that have different identifiers in their labels.

labeled statistical channel. In integrated services digital networks (ISDN), a labeled channel in which the block payloads or the duration of each successive block is random.

Lambertian radiator. *See* Lambert's cosine law.

Lambertian reflector. *See* Lambert's cosine law.

Lambertian source. An optical source that obeys Lambert's cosine law. *Note:* Conventional (surface-emitting) LEDs approximate a Lambertian source. They have a large beam divergence, and a radiation pattern that approximates a sphere. Thus, most of their total optical output is not coupled into communications fibers.

Lambert's cosine law. The mathematical statement that the radiance of certain idealized optical sources is directly proportional to the cosine of the angle—with respect to the direction of maximum radiance—from which the source is viewed. *Note:* Lambert's cosine law may also apply to certain idealized diffuse reflectors. *Synonyms* cosine emission law, Lambert's emission law.

Lambert's emission law. *Synonym* Lambert's cosine law.

LAN. *Acronym for* **loc**al area network.

LAN application (software). An application software package specifically designed to operate in a local-area-network environment.

land line. A colloquial name for conventional telephone facilities. *Note:* Land lines include conventional twisted-pair lines, carrier facilities, and microwave radio facilities for supporting a conventional telephone channel, but do not include satellite links or mobile telephone links using radio transmissions.

land mobile-satellite service. A mobile-satellite service in which mobile Earth stations are located on land.

land mobile service. A mobile service between base stations and land mobile stations, or between land mobile stations.

land mobile station. A mobile station in the land mobile service capable of surface movement within the geographical limits of a country or continent.

landscape mode. 1. In facsimile, the mode for scanning lines across the longer dimension of a rectangular object, *i.e.*, rectangular original. 2. In computer graphics, the orientation of an image in which the longer dimension is horizontal. 3. An orientation of printed text on a page such that the lines of text are parallel to the long dimension of the page. *Note:* If the page contains an image, such as a picture, and the page is viewed in the normal manner, the long dimension of the page would be parallel to the line that joins the eyes of the viewer.

land station. A station in the mobile service not intended to be used while in motion.

language. A set of characters, conventions, and rules that is used for conveying information.

language processor. A program that performs tasks, such as translating and interpreting, required for processing a specified programming language. *Note:* Examples of language processors include a Fortran processor and a COBOL processor.

LAN operating system *See* network operating system.

LAP-B. The Data Link Layer protocol as specified by CCITT Recommendation X.25 (1989).

LAP-D. *Abbreviation for* link access procedure D. A link protocol used in ISDN.

laser. *Acronym for* light amplification by stimulated emission of radiation. A device that produces a coherent beam of optical radiation by stimulating electronic, ionic, or molecular transitions to higher energy levels so that when they return to lower energy levels they emit energy. *Note 1:* Laser radiation may be either temporally coherent, spatially coherent, or both. *Note 2:* The degree of coherence of laser radiation exceeds 0.88.

laser chirp. An abrupt change of the center wavelength of a laser, caused by laser instability.

laser diode. *Synonym* injection laser diode.

laser disk. *See* optical disk.

laser intelligence (LASINT). Technical and geolocation intelligence derived from laser systems; a subcategory of electro-optical intelligence.

laser medium. *Synonym* active laser medium.

lasing. *See* laser.

lasing threshold. The lowest excitation level at which laser output is dominated by stimulated emission rather than by spontaneous emission.

LASINT. *Acronym for* laser intelligence.

last-in first-out (LIFO). A queuing discipline in which entities in a queue leave in the reverse order of the sequence in which they arrive. *Note:* Service, when available, is offered to the entity that has been in the LIFO queue the shortest time.

LATA. *Acronym for* local access and transport area.

lateral offset loss. 1. In fiber optics, a loss of optical power at a splice or connector, caused by a lateral, *i.e.*, transverse, offset of the mating fiber cores, which offset causes an imperfect transfer of the optical signal from the "transmitting" fiber to the "receiving" fiber. *Note:* The effect of a given amount of lateral offset will depend on other parameters such as the relative diameters of the respective cores. For example: if, because of manufacturing tolerances, the "transmitting" core is smaller than the "receiving" core, the effect will be less than if both cores were the same size. 2. An analogous loss of optical power caused by lateral misalignment of the fiber and optical source. *Synonym* transverse offset loss.

launch angle. 1. The angle, with respect to the normal, at which a light ray emerges from a surface. 2. The beam divergence at an emitting surface, such as that of a light-emitting diode (LED), laser, lens, prism, or optical fiber end face. 3. At an end face of an optical fiber, the angle between an input ray and the fiber axis. *Note:* If the end face of the fiber is perpendicular to the fiber axis, the launch angle is equal to the incidence angle when the ray is external to the fiber and the refraction angle when initially inside the fiber.

launching fiber. An optical fiber used in conjunction with a source to excite the modes of another fiber in a particular fashion. *Note:* Launching fibers are most often used in test systems to improve the precision of measurements. *Synonym* injection fiber.

launch numerical aperture (LNA). The numerical aperture of an optical system used to couple (launch) power into an optical fiber. *Note 1:* LNA may differ from the stated NA of a final focusing element if, for example, that element is underfilled or the focus is other than that for which the element is specified. *Note 2:* LNA is one of the parameters that determine the initial distribution of power among the modes of an optical fiber.

layer. 1. In radio wave propagation, *see* F region. 2. In telecommunications networks and open systems architecture, a group of related functions that are performed in a given level in a hierarchy of groups of related functions. *Note:* In specifying the functions for a given layer, the assumption is made that the specified functions for the layers below are performed, except for the lowest layer.

layered system. A system in which components are grouped, *i.e.*, layered, in a hierarchical arrangement, such that lower layers provide functions and services that support the functions and services of higher layers. *Note:* Systems of ever-increasing complexity and capability can be built by adding or changing the layers to improve overall system capability while using the components that are still in place.

lay length. In communications cables—including fiber-optic cables—having the transmission media wrapped helically around a central member, the longitudinal distance along the cable required for one complete helical wrap; *i.e.*, the total cable length divided by the total number of wraps. *Note 1:* In many fiber-optic cable designs, the lay length is shorter than in metallic cables of similar diameter, to avoid overstressing the fibers during the pulling associated with the installation operation. *Note 2:* The wraps, *i.e.*, turns, that are referred to should not be confused with the twists given twisted metallic pairs, *i.e.*, wires, to reduce electromagnetic coupling. Pairs of optical fibers are not given such twists. *Synonym* pitch.

LBO. *Abbreviation for* **line buildout.** *Synonym* building out.

LCD. *Abbreviation for* liquid crystal display.

LDM. *Abbreviation for* limited distance modem.

leaky bucket counter. A counter that is incremented by unity each time an event occurs and that is periodically decremented by a fixed value.

leaky mode. In an optical fiber, a mode having a field that decays monotonically for a finite distance in the transverse direction but becomes oscillatory everywhere beyond that finite distance. *Note:* Leaky modes correspond to leaky rays in the terminology of geometric optics. Leaky modes experience attenuation, even if the waveguide is perfect in every respect. *Synonym* tunneling mode.

leaky ray. In an optical fiber, a ray for which geometric optics would predict total internal reflection at the boundary between the core and the cladding, but which suffers loss by virtue of the curved core boundary. *Note:* Leaky rays correspond to leaky (*i.e.*, tunneling) modes in the terminology of mode descriptors. *Synonym* tunneling ray.

leap second. An occasional adjustment of one second, added to, or subtracted from, Coordinated Universal Time (UTC) to bring it into approximate synchronism with UT-1, which is the time scale based on the rotation of the Earth. *Note 1:* Adjustments, when required, are made at the end of June 30, or preferably, December 31, Universal Time, so that UTC never deviates from UT-1 by more than 0.9 second. *Note 2:* The last minute of the day on which an adjustment is made has 61 or 59 seconds.

leased circuit. Dedicated common-carrier facilities and channel equipment used by a network to furnish exclusive private line service to a specific user or group of users.

least-time principle. *Synonym* Fermat's principle.

LEC. *Abbreviation for* local exchange carrier.

LED. *Abbreviation for* light-emitting diode.

left-hand (anti-clockwise) polarized wave. An elliptically or circularly polarized wave, in which the electric field vector, observed in the fixed plane, normal to the direction of propagation, whilst looking in the direction of propagation, rotates with time in a left-hand or anticlockwise direction.

leg. 1. A segment of an end-to-end route or path, such as a path from user to user via several networks and nodes within networks. *Note:* Examples of legs are several sequential microwave links between two switching centers and a transoceanic cable between two shore communications facilities, each connected to a node in a national network. 2. A connection from a specific node to an addressable entity, such as communication link from a computer workstation to a hub.

level. 1. The absolute or relative voltage, current, or power at a particular point in a circuit or system. 2. A tier or layer of a hierarchical system, *e.g.*, the Link-Level protocol, high-level computer language.

level alignment. The adjustment of transmission levels of single links and of links in tandem to prevent problems such as overloading of transmission subsystems.

LF. *Abbreviation for* low frequency.

LIFO. *Acronym for* last-in first-out.

light. In a strict sense, the region of the electromagnetic spectrum that can be perceived by human vision, *i.e.*, the visible spectrum, which is approximately the wavelength range of 0.4 μm to 0.7 μm. *Note 1:* In the laser and optical communications fields, custom and practice have extended usage of the term *light* to include the much broader portion of the electromagnetic spectrum that can be handled by the basic optical techniques used for the visible spectrum. *Note 2:* The region embraced by the term *light* has not been clearly defined, but by convention and usage, is considered to extend from the near-ultraviolet region of approximately 0.3 μm, through the visible region, and into the mid-infrared region to approximately 30 μm.

light-emitting diode (LED). A semiconductor device that emits incoherent optical radiation when biased in the forward direction.

lightguide. *See* optical fiber.

lightning down-conductor. In a lightning protection subsystem, the conductor connecting the air terminal or overhead ground wire to the earth electrode subsystem.

lightning protection subsystem. All of the components used to protect a facility from the effects of lightning. *Note:* The lightning protection subsystem includes air terminals, lightning down- conductors, the earth electrode subsystem, air gaps, arresters, and their interconnections.

light pen. A stylus, usually hand-held, that contains a photodetector or light source, and that allows interaction with a computer through a specially designed monitor screen.

light valve. *Synonym* optical switch.

limited protection. A form of short-term communications security applied to the electromagnetic or acoustic transmission of unclassified information that warrants a degree of protection against simple analysis and easy exploitation but that does not warrant protection to the extent needed for security of classified information.

limited-protection voice equipment. Equipment that provides limited security for unclassified voice communications.

limiter. A device in which the voltage or some other characteristic of the output signal is automatically prevented from exceeding a specified value.

limiter circuit. A circuit of nonlinear elements that restricts the electrical excursion of a variable in accordance with specified criteria.

limiting. Any process by which a specified characteristic (usually amplitude) of the output of a device is prevented from exceeding a predetermined value. *Note 1: Hard limiting ("clipping")* is a limiting action in which there is (a) over the permitted dynamic range, negligible variation in the expected characteristic of the output signal, and (b) a steady-state signal, at the maximum permitted level, for the duration of each period when the output would otherwise be required to exceed the permitted dynamic range in order to correspond to the transfer function of the device. *Note 2: Soft limiting* is limiting in which the transfer function of a device is a function of its instantaneous or integrated output level. The output waveform is therefore distorted, but not clipped.

limits of interference. In radio transmission, the maximum permissible interference as specified in recommendations of the International Special Committee on Radio Interference or other recognized authority.

line. 1. A physical medium for transferring electrical or electromagnetic energy from one point to another for purposes of communications. 2. A land line. 3. A metallic medium used for the transmission of electrical power. 4. *See* scanning line.

line adapter circuit. *See* four-wire terminating set.

linear analog control. *Synonym* linear analog synchronization.

linear analog synchronization. Synchronization in which the functional relationships used to obtain synchronization are of simple proportionality. *Synonym* linear analog control.

linear combiner. A diversity combiner in which the combining consists of simple addition of two or more signals.

linear device. A device for which the output is, within a given dynamic range, linearly proportional to the input.

linearity. The property of a system in which, if input signals X and Y result in system output $S(X)$ and $S(Y)$ respectively, the input signal $aX + bY$ will result in the output $aS(X) + bS(Y)$, where S is the system transfer function and a and b are scalars.

linearly polarized (LP) mode. A mode for which the field components in the direction of propagation are small compared to components perpendicular to that direction. *Note:* The LP description is an approximation that is valid for a weakly guiding optical fiber, including typical telecommunications grade fibers.

linear network. *See* network topology.

linear optimization. *Synonym* linear programming.

linear polarization. Of an electromagnetic wave, confinement of the E-field vector or H-field vector to a given plane. *Note:* Historically, the orientation of a polarized electromagnetic wave has been defined in the optical regime by the orientation of the electric vector, and in the radio regime, by the orientation of the magnetic vector. *Synonym* plane polarization.

linear predictive coding (LPC). A method of digitally encoding analog signals, which method uses a single-level or multilevel sampling system in which the value of the signal at each sample time is predicted to be a linear function of the past values of the quantized signal. *Note:* LPC is related to adaptive predictive coding (APC) in that both use adaptive predictors. However, LPC uses more prediction coefficients to permit use of a lower information bit rate than APC, and thus requires a more complex processor.

linear programming (LP). In operations research, a procedure for locating the maximum or minimum of a linear function of variables that are subject to linear constraints. *Synonym* linear optimization.

linear topology. *See* network topology.

line balance. The degree of electrical similarity of the two conductors of a transmission line. *Note:* A high degree of line balance reduces pickup of extraneous disturbances of all kinds, including crosstalk.

line buildout (LBO). *Synonym* building out.

line code. A code chosen for use within a communications system for transmission purposes. *Note 1:* A line code may differ from the code generated at a user terminal, and thus may require translation. *Note 2:* A line code may, for example, reflect a requirement of the transmission medium, *e.g.*, optical fiber versus shielded twisted pair.

line driver. An amplifier used to enhance the transmission reliability of a usually digital intrafacility metallic transmission line, over extended distances, by driving the input to the transmission line with a higher than normal signal level. *Note:* An example of a line driver is an amplifier used to extend the range of an RS-232C digital signal beyond 50 feet (~15 m) while maintaining a specified bit-error ratio.

line filter balance. A network designed to maintain phantom group balance when one side of the group is equipped with a carrier system. *Note:* Since it must balance the phantom group for only voice frequencies, the line filter balance configuration is usually simple compared with the filter that it balances.

line hit. *See* hit.

line load control. A network-provided service feature that allows selective denial of call origination to certain lines when excessive demands for service are required of a switching center.

line loop. *See* loop.

line-of-sight (LOS) propagation. Of an electromagnetic wave, propagation in which the direct ray from the transmitter to the receiver is unobstructed, *i.e.,* the transmission path is not established by or dependent upon reflection or diffraction. *Note:* The need for LOS propagation is most critical at VHF and higher frequencies.

line-route map. A map or overlay for signal communications operations that shows the actual routes and types of construction of wire circuits in the field. It also gives the locations of switchboards and telegraph stations.

line side. The portion of a device that is connected to external, *i.e.,* outside plant, facilities such as trunks, local loops, and channels.

line source. 1. In spectroscopy, an optical source that emits one or more spectrally narrow lines as opposed to a continuous spectrum. 2. In the geometric sense, an optical source having an emitting area in the form of a spatially narrow line, *e.g.*, a slit. *Synonym* slit source.

line spectrum. In optics, an emission or absorption spectrum consisting of one or more narrow spectral lines, as opposed to a continuous spectrum.

line speed. *See* modulation rate.

line-to-line correlation. In facsimile, the correlation of object information from scanning line to scanning line. *Note:* Line-to-line correlation is used in two-dimensional encoding.

line traffic coordinator (LTC). In a DDN switching center, the processor that controls traffic on a line.

line verification. *See* busy verification.

linewidth. *See* spectral width.

link. 1. The communications facilities between adjacent nodes of a network. 2. A portion of a circuit connected in tandem with, *i.e.*, in series with, other portions. 3. A radio path between two points, called a radio link. 4. In communications, a general term used to indicate the existence of communications facilities between two points. 5. A conceptual circuit, *i.e.*, logical circuit, between two users of a network, that enables the users to communicate, even when different physical paths are used. *Note 1:* In all cases, the type of link, such as data link, downlink, duplex link, fiber optic link, line-of-sight link, point-to-point link, radio link and satellite link, should be identified. *Note 2:* A link may be simplex, half-duplex, or duplex. 6. In a computer program, a part, such as a single instruction or address, that passes control and parameters between separate portions of the program. 7. In hypertext, the logical connection between discrete units of data.

link encryption. The application of on-line crypto-operation to a link of a communications system so that all information passing over the link is encrypted in its entirety.

linking protection (LP). In adaptive high-frequency (HF) radio, protection intended to prevent the establishment of unauthorized links or the unauthorized manipulation of legitimate links, and which are administered through an authorization process.

Link Layer. *Deprecated term for* Data Link Layer. *See* Open Systems Interconnection—Reference Model.

link level. In the hierarchical structure of a primary or secondary station, the conceptual level of control or data processing logic that controls the data link. *Note:* Link-level functions provide an interface between the station high-level logic and the data link. Link-level functions include (a) transmit bit injection and receive bit extraction, (b) address and control field interpretation, (c) command response generation, transmission and interpretation, and (d) frame check sequence computation and interpretation.

link orderwire. A voice or data communications circuit that (a) serves as a transmission link between adjacent communications facilities that are interconnected by a transmission link and (b) is used only for coordination and control of link activities, such as traffic monitoring and traffic control.

link protocol. A set of rules relating to data communications over a data link. *Note:* Link protocols define data link parameters, such as transmission code, transmission mode, control procedures, and recovery procedures.

link quality analysis (LQA). In adaptive high-frequency (HF) radio, the overall process by which measurements of signal quality are made, assessed, and analyzed. *Note 1:* In LQA, signal quality is determined by measuring, assessing, and analyzing link parameters, such as bit error ratio (BER), and the levels of the ratio of signal-plus-noise-plus-distortion to noise-plus-distortion (SINAD). Measurements are stored at—and exchanged between—stations, for use in making decisions about link establishment. *Note 2:* For adaptive HF radio, LQA is automatically performed and is usually based on analyses of pseudo-BERs and SINAD readings.

lip synchronization. The synchronization of audio and corresponding video signals so that there is no noticeable lack of simultaneity between them. *Note:* An example of a lip synchronization problem is the case in which television video and audio signals are transported via different facilities *(e.g.,* a geosynchronous satellite link and a landline) that have significantly differently delay times, respectively. In such cases it is necessary to delay the audio electronically to allow for the difference in propagation times.

liquid crystal display (LCD). A display device that creates characters by means of the action of electrical signals on a matrix of liquid cells that become opaque when energized. *Note:* A liquid crystal display may be designed to be viewed by reflected or transmitted light.

LLC. *Abbreviation for* logical link control. *See* logical link control sublayer.

LNA. *Abbreviation for* launch numerical aperture.

load. 1. The power consumed by a device or circuit in performing its function. 2. A power-consuming device connected to a circuit. 3. To enter data or programs into storage or working registers. 4. To insert data values into a database that previously contained no occurrences of data. 5. To place a magnetic tape reel on a tape drive, or to place cards into the card hopper of a card punch or reader.

load capacity. In pulse-code modulation (PCM), the level of a sinusoidal signal that has positive and negative peaks that coincide with the positive and negative virtual decision values of the encoder. *Note:* Load capacity is usually expressed in dBm0. *Synonym* overload point.

loader. A routine that reads data into main storage.

load factor. The ratio of the average load over a designated period of time to the peak load occurring during that period.

loading. 1. The insertion of impedance into a circuit to change the characteristics of the circuit. 2. In multichannel communications systems, the insertion of white noise or equivalent dummy traffic at a specified level to simulate system traffic and thus enable analysis of system performance. 3. In telephone systems, the load, *i.e.*, power level, imposed by the busy hour traffic. *Note 1:* The loading may be expressed as (a) the equivalent mean power and the peak power as a function of the number of voice channels or (b) the equivalent power of a multichannel complex or signal composite referred to zero transmission level point (0TLP). *Note 2:* Loading is a function of the number of channels and the specified voice channel mean power.

loading characteristic. In multichannel telephone systems, a plot, for the busy hour, of the equivalent mean power and the peak power as a function of the number of voice channels. *Note:* The equivalent power of a multichannel signal referred to the zero transmission level point is a function of the number of channels and has for its basis a specified voice channel mean power.

loading coil. A coil that does not provide coupling to any other circuit, but is inserted in a circuit to increase its inductance. *Note 1:* Loading coils inserted periodically in a pair of wires reduce the attenuation at the higher voice frequencies up to the cutoff frequency of the low-pass filter formed by (a) the inductance of the coils and distributed inductance of the wires, and (b) the distributed capacitance between the wires. Above the cutoff frequency, attenuation increases rapidly. *Note 2:* A common application of loading coils is to improve the voice-frequency amplitude response characteristics of twisted cable pairs. When connected across a twisted pair at regular intervals, loading coils, in concert with the distributed resistance and capacitance of the pair, form an audio-frequency filter that improves the high-frequency audio response of the pair. *Note 3:* When loading coils are in place, signal attenuation increases rapidly for frequencies above the audio cutoff frequency. Thus, when a pair is used to support applications that require higher frequencies, such as carrier systems, loading coils must be absent.

lobe. 1. An identifiable segment of an antenna radiation pattern. *Note:* A lobe is characterized by a localized maximum bounded by identifiable nulls. 2. A pair of channels between a data station and a lobe attaching unit, one channel for sending and one for receiving, as seen from the point of view of the attached data station.

lobe attaching unit. In a ring network, a functional unit used to connect and disconnect data stations to and from the ring without disrupting network operations.

local access and transport area (LATA). Under the terms of the Modification of Final Judgment (MFJ), a geographical area within which a divested Bell Operating Company (BOC) is permitted to offer exchange telecommunications and exchange access services. *Note:* Under the terms of the MFJ, the BOCs are generally prohibited from providing services that originate in one LATA and terminate in another.

local area network (LAN). A data communications system that (a) lies within a limited spatial area, (b) has a specific user group, (c) has a specific topology, and (d) is not a public switched telecommunications network, but may be connected to one. *Note 1:* LANs are usually restricted to relatively small areas, such as rooms, buildings, ships, and aircraft. *Note 2:* An interconnection of LANs within a limited geographical area, such as a military base, is commonly referred to as a campus area network. An interconnection of LANs over a city-wide geographical area is commonly called a metropolitan area network (MAN). An interconnection of LANs over large geographical areas, such as nationwide, is commonly called a wide area network (WAN). *Note 3:* LANs are not subject to public telecommunications regulations.

local battery. 1. In telegraphy, the source of power that actuates the telegraphic station recording instruments, as distinguished from the source of power that furnishes current to the line. 2. In telephony, a system in which each telephone instrument has its own source of power, as opposed to being powered from the central office. 3. A source of local power for a telephone instrument.

local call. 1. Any call using a single switching facility. 2. Any call for which an additional charge, *i.e.*, toll charge, is not made to the calling or called party. *Note:* Calls such as those via "800" numbers do not qualify as local calls, because the called party is charged.

local central office. *Synonym* central office.

local clock. A source of timing located in close proximity to an associated facility, such as a communications station, central office, or node. *Note:* The same clock might be a remote clock relative to some other facility.

local exchange. *Synonym* central office.

local exchange carrier (LEC). A local telephone company, *i.e.*, a communications common carrier that provides ordinary local voice-grade telecommunications service under regulation within a specified service area.

local exchange loop. An interconnection between customer premises equipment and telephone central office.

local line. *See* loop.

local loop. *Synonym* loop.

local measured service. *See* measured-rate service.

local office. *Synonym* central office.

local orderwire. A communications circuit between a technical control facility and selected terminal or repeater locations within the communications complex. *Note:* In multichannel radio systems, the local orderwire is usually a handset connection at the radio location.

local side. The portion of a device that is connected to internal facilities, such as switches, patch panels, test bays and supervisory equipment.

lock-in frequency. A frequency at which a closed-loop system can acquire and track a signal. *See* lock-in range.

lock-in range. 1. The range of frequencies within which a closed-loop system can aquire and track a signal. 2. The dynamic range within which a closed-loop system can aquire and track a signal.

lockout. 1. In telephone systems, treatment of a user's line or trunk that is in trouble, or in a permanent off-hook condition, by automatically disconnecting the line from the switching equipment. 2. In public telephone systems, a process that denies an attendant or other users the ability to reenter an established connection. 3. In a telephone circuit controlled by two voice-operated devices, the inability of one or both users to get through, either because of excessive local circuit noise or because of continuous speech from either or both users. 4. In mobile communications, an arrangement of control circuits whereby only one receiver can feed the system at a time. *Synonym* receiver lockout system. 5. An arrangement for restricting access to use of all, or part of, a computer system. *Synonym* protection.

log. *Synonym* journal.

logical circuit. *Synonym* virtual circuit.

logical link control (LLC) sublayer. In a local-area-network/metropolitan-area-network (LAN/MAN) system, the part of the link level that (a) supports medium-independent data link functions and (b) uses the services of the medium access control sublayer to provide services to the network layer.

logical route. *Synonym* virtual circuit.

logical signaling channel. A logical channel that provides a signaling path within an information channel or within a physical signaling channel.

logical topology. Of a network, the schematic configuration that reflects the network's function, use, or implementation without regard to the physical interconnection of network elements.

log in. To perform a login procedure. *Synonym* log on.

login. The procedure that is followed by a user in beginning a session, *e.g.*, a period of terminal operation. *Synonym* logon.

log off. To perform a log-off procedure. *Synonym* log out.

log-off. The procedure that is followed by a user in closing a session, *e.g.*, a period of terminal operation. *Synonym* log-out.

logon. *Synonym* login.

log on. *Synonym* log in.

log out. *Synonym* log off.

log-periodic (LP) antenna. A broadband, multielement, unidirectional, narrow-beam antenna that has impedance and radiation characteristics that are regularly repetitive as a logarithmic function of the excitation frequency. *Note:* The length and spacing of the elements of a log-periodic antenna increase logarithmically from one end to the other. *Synonym* log-periodic array.

log-periodic (LP) array. *Synonym* log-periodic antenna.

long-distance call. Any telephone call to a destination outside the local service area of the calling station, whether inter-LATA or intra-LATA, and for which there is a charge beyond that for basic service. *Synonym* toll call.

long-haul communications. 1. In public switched networks, pertaining to circuits that span large distances, such as the circuits in inter-LATA, interstate, and international communications. 2. In the military community, communications among users on a national or worldwide basis. *Note 1:* Compared to tactical communications, long-haul communications are characterized by (a) higher levels of users, such as the National Command Authority, (b) more stringent performance requirements, such as higher quality circuits, (c) longer distances between users, including world wide distances, (d) higher traffic volumes and densities, (e) larger switches and trunk cross sections, and (f) fixed and recoverable assets. *Note 2: "Long-haul communications"* usually pertains to the U.S. Defense Communications System.

longitudinal balance. 1. The electrical symmetry, with respect to ground, of the two wires of a pair 2. An expression of the difference in impedance of the two sides of a circuit.

longitudinal offset loss. *Synonym* gap loss.

longitudinal redundancy check (LRC). A system of error control based on the formation of a block check following preset rules. *Note 1:* The block check formation rules are applied in the same manner to each character. *Note 2:* A combination of longitudinal and vertical redundancy check allows the detection and correction of single bit errors. *Synonym* horizontal redundancy check.

longitudinal voltage. A voltage induced or appearing along the length of a transmission medium. *Note 1:* Longitudinal voltage may be effectively eliminated by using differential amplifiers or receivers that respond only to voltage differences, *e.g.,* those between the wires that constitute a pair. *Note 2:* Induced longitudinal voltages at low (power-line) frequencies can be greatly reduced by twisting parallel wires to create what are referred to as *"twisted wire pairs."* *Synonym* common-mode voltage.

long line. A transmission line in a long-distance communications network. *Note:* Examples of long lines are TDM and FDM carrier systems, microwave radio links, geosynchronous satellite links, underground cables, aerial cables and open wire, and submarine cables.

long-range aid to navigation (loran) system. *See* loran.

long-range radio aid to navigation system. *See* loran.

long-term stability. Of an oscillator, the degree of uniformity of frequency over time, when the frequency is measured under identical environmental conditions, such as supply voltage, load, and temperature. *Note:* Long-term frequency changes are caused by changes in the oscillator elements that determine frequency, such as crystal drift, inductance changes, and capacitance changes.

long wavelength. In fiber optic communications, pertaining to optical wavelengths greater than = 1 μm.

look-ahead-for-busy (LFB) information. Information concerning network resources available to support higher precedence calls. *Note 1:* Available resources include idle circuits and circuits used for lower precedence calls. *Note 2:* LFB information may be used to make call-path reservations.

loop. 1. A communications channel from a switching center or an individual message distribution point to the user terminal. *Synonym* subscriber line. 2. In telephone systems, a pair of wires from a central office to a subscriber's telephone. *Synonyms* local loop, user line. 3. Go-and-return conductors of an electric circuit; a closed circuit. 4. A closed path under measurement in a resistance test. 5. A type of antenna, in the form of a circle or rectangle, usually used in direction-finding equipment and in UHF reception. 6. A sequence of instructions that may be executed iteratively while a certain condition prevails until the loop has been executed once.

loop-back. 1. A method of performing transmission tests of access lines from the serving switching center, which method usually does not require the assistance of personnel at the served terminal. 2. A method of testing between stations (not necessarily adjacent) wherein two lines are used, with the testing being done at one station and the two lines interconnected at the distant station. 3. A patch, applied manually or automatically, remotely or locally, that facilitates a loop-back test.

loop check. *Synonym* echo check.

looped dual bus. A distributed-queue dual-bus (DQDB) scheme in which the head-of-bus functions for both buses are at the same location.

loop filter. In a phase-locked loop, a filter located between the phase detector (or time discriminator) and the voltage controlled oscillator (or phase shifter).

loop gain. 1. The total usable power gain of a carrier terminal or two-wire repeater. *Note:* The maximum usable gain is determined by, and may not exceed, the losses in the closed path. 2. The sum of the gains, expressed in dB, acting on a signal passing around a closed path, *i.e.*, a loop.

loop noise. The noise contributed by one or both loops of a telephone circuit to the total circuit noise. *Note:* In a given case, it should be stated whether the loop noise is for one or both loops.

loop start. A supervisory signal given by a telephone or PBX in response to completing the loop path.

loop test. A test that uses a closed circuit, *i.e.*, loop, to detect and locate faults.

loop transmission. Multipoint transmission in which (a) all the stations in a network are serially connected in one closed loop, (b) there are no cross-connections, (c) the stations serve as regenerative repeaters, forwarding messages around the loop until they arrive at their destination stations, and (d) any station can introduce a message into the loop by interleaving it with other messages.

loose buffer. *See* buffer.

loran. *Acronym for* long-range radio navigation. A long-range radio navigation position-fixing system consisting of an array of fixed stations that transmit precisely synchronized signals to mobile receivers. *Note:* A loran receiver measures differences in the times of arrival of the signals from the various stations. A fixed difference in the time of arrival of the signals from any two stations will define a hyperbolic arc on which the receiver must lie. Three or more stations are needed to remove ambiguities in the position of the receiver. *Synonyms* long-range aid to navigation system, long-range radio aid to navigation system.

loran station. A long distance radionavigation land station transmitting synchronized pulses. Hyperbolic lines of position are determined by the measurement of the difference in the time of arrival of these pulses.

LOS. *Abbreviation for* line of sight. *See* line-of-sight propagation.

loss. 1. The diminution, usually expressed in dB, of signal level in a communications medium. 2. The power, usually expressed in watts, consumed by a circuit or component. 3. The energy dissipated without accomplishing useful work or purpose.

lossy medium. A medium in which a significant amount of the energy of a propagating electromagnetic wave is absorbed per unit distance traveled by the wave.

lost block. A block not delivered to the user within a specified maximum end-to-end block transfer time.

lost call. A call that has not been completed for any reason other than cases where the call receiver (termination) is busy.

lower frequency ground. *Deprecated term. See* facility grounding system.

lowest usable high frequency (LUF). The lowest frequency in the HF band at which the received field intensity is sufficient to provide the required signal-to-noise ratio for a specified time period, *e.g.*, 0100 to 0200 UTC, on 90% of the undisturbed days of the month.

low frequency (LF). Any frequency in the band from 30 kHz to 300 kHz.

low-level keying. *Synonym* low-level signaling.

low-level language. *Synonym* computer-oriented language.

low-level modulation. Modulation of a signal, *e.g.,* a carrier, at a point in a system or device, such as a radio transmitter, where the power level is low compared to the final output power.

low-level signaling. The use on signal lines of voltage levels that are between the limits of positive and negative 6 volts. *Synonym* low-level keying.

low-pass filter. A filter network that passes all frequencies below a specified frequency with little or no loss, but strongly attenuates higher frequencies.

low-performance equipment. 1. Equipment that has imprecise characteristics that do not meet system reliability requirements. 2. In military communications, equipment that has insufficiently exacting characteristics to permit its use in trunks or links. *Note:* Low-performance equipment may be used in loops if it meets loop performance requirements. 3. Tactical ground and airborne equipment that (a) has size, weight, and complexity characteristics that must be kept to a minimum and (b) is used in systems that have components with similar minimum performance characteristics.

low-power communication device. A restricted radiation device, exclusive of those employing conducted or guided radio frequency techniques, used for the transmission of signs, signals (including control signals), writing, images and sounds or intelligence of any nature by radiation of electromagnetic energy. Examples: Wireless microphone, phonograph oscillator, radio-controlled garage door opener, and radio-controlled models.

LP. *Abbreviation for* linear programming, linking protection.

L-pad. A pad composed of two discrete components, one series component and one shunt component. *Note:* In schematic representation, the components resemble the upper-case letter "L,"hence the name.

LPC. *Abbreviation for* linear predictive coding.

LP mode. *Abbreviation for* linearly polarized mode.

LP_{01} mode. Designation of the fundamental LP mode. *See* fundamental mode.

LQA. *Abbreviation for* link quality analysis.

LRC. *Abbreviation for* longitudinal redundancy check.

LSB. *Abbreviation for* least significant bit, lower sideband.

LTC. *Abbreviation for* line traffic coordinator.

LUF. *Acronym for* lowest usable high frequency.

luminescent diode. *See* superluminescent LED.

lynx. A World Wide Web browser that provides a character-based user interface to hypertext-based information. *Note:* Lynx can display only character-based portions of the hypertext-based information.

MAC. *Abbreviation for* medium access control. *See* medium access control sublayer.

machine-independent. In telecommunications, computer, and data processing systems, pertaining to operations, procedures, computer programs, and processing that do not depend upon specific hardware for their successful execution.

machine instruction. An instruction that is written in a machine language and can be executed directly by the processor for which it was designed without translation or interpretation.

machine language. A language that need not be modified, translated, or interpreted before it can be used by the processor for which it was designed. *Note 1:* The operation codes and addresses used in instructions written in machine language can be directly sensed by the arithmetic and control unit circuits of the processor for which the language is designed. *Note 2:* Instructions written in an assembly language or a high-level language must be translated into machine language before they can be executed by a processor. *Note 3:* Machine languages are usually used by computer designers rather than computer users.

machine learning. The ability of a device to improve its performance based on its past performance.

machine-oriented language. *Synonym* computer-oriented language.

machine-readable medium. A medium capable of storing data in a form that can be accessed by an automated sensing device. *Note:* Examples of machine-readable media include (a) magnetic disks, cards, tapes, and drums, (b) punched cards and paper tapes, (c) optical disks, and (d) magnetic ink characters. *Synonym* automated data medium.

machine word. *Synonym* computer word.

macrobend. A relatively large-radius bend in an optical fiber, such as might be found in a splice organizer tray or a fiber-optic cable that has been bent. *Note:* A macrobend will result in no significant radiation loss if it is of sufficiently large radius. The definition of "sufficiently large" depends on the type of fiber. Single-mode fibers have a low numerical aperture, typically less than 0.15, and are therefore are more susceptible to bend losses than other types. Normally, they will not tolerate a minimum bend radius of less than 6.5 to 7.5 cm (2.5 to 3 inches). Certain specialized types of single-mode fibers, however, can tolerate a far shorter minimum bend radius without appreciable loss. A graded-index multimode fiber having a core diameter of 50 µm and a numerical aperture of 0.20 will typically tolerate a minimum bend radius of not less than 3.8 cm (1.5 inches). The fibers commonly used in customer-premises applications (62.5-µm core) typically have a relatively high numerical aperture, (approximately 0.27), and can tolerate a bend radius of less than an inch (2.5 cm).

macrobend loss. In an optical fiber, that loss attributable to macrobending. *Synonym* curvature loss.

magnetic card. A card with a magnetizable surface on which data can be stored and retrieved.

magnetic circuit. 1. The complete closed path taken by magnetic flux. 2. A region of ferromagnetic material, such as the core of a transformer or solenoid, that contains essentially all of the magnetic flux.

magnetic core storage. In computer technology, a storage device that uses ferromagnetic materials such as iron, iron oxide, or ferrite and in such shapes as wires, toroids, and rods.

magnetic disk. *See* diskette, hard disk.

magnetic drum. A right circular cylinder with a magnetizable surface on which digital data can be stored and retrieved.

magnetic storm. A perturbation of the Earth's magnetic field, caused by solar disturbances, usually lasting for a brief period (several days) and characterized by large deviations from the usual value of at least one component of the field. *Note:* Magnetic storms can affect radio propagation because they disturb the ionosphere.

magnetic tape. 1. A tape with a magnetizable surface on which data can be stored and retrieved. 2. A tape or ribbon of any material impregnated or coated with magnetic or other material on which information may be placed in the form of magnetically polarized spots.

magneto-ionic double refraction. The combined effect of the Earth's magnetic field and atmospheric ionization, whereby a linearly polarized wave entering the ionosphere is split into two components called the ordinary wave and the extraordinary wave. *Note:* The component waves follow different paths, experience different attenuations, have different phase velocities, and, in general, are elliptically polarized in opposite senses.

magneto-optic. *See* magneto-optic effect.

magneto-optic effect. Any one of a number of phenomena in which an electromagnetic wave interacts with a magnetic field, or with matter under the influence of a magnetic field. *Note:* The most important magneto-optic effect having application to optical communication is the Faraday effect, in which the plane of polarization is rotated under the influence of a magnetic field parallel to the direction of propagation. This effect may be used to modulate a lightwave.

mailbox-type facility. A facility in which a message from an originating user is stored until the destination user requests delivery of that message.

main beam. *Synonym* main lobe.

main distribution frame (MDF). A distribution frame on one part of which the external trunk cables entering a facility terminate, and on another part of which the internal user subscriber lines and trunk cabling to any intermediate distribution frames terminate. *Note 1:* The MDF is used to cross-connect any outside line with any desired terminal of the multiple cabling or any other outside line. *Note 2:* The MDF usually holds central office protective devices and functions as a test point between a line and the office. *Note 3:* The MDF in a private exchange performs functions similar to those performed by the MDF in a central office. *Synonym (in telephony)* **main frame.**

main frame. *Synonym (in telephony)* main distribution frame.

mainframe. A large computer, usually one to which other computers and/or terminals are connected to share its resources and computing power.

main lobe. Of an antenna radiation pattern, the lobe containing the maximum power (exhibiting the greatest field strength). *Note:* The horizontal radiation pattern, *i.e.*, that which is plotted as a function of azimuth about the antenna, is usually specified. The width of the main lobe is usually specified as the angle encompassed between the points where the power has fallen 3 dB below the maximum value. The vertical radiation pattern, *i.e.*, that which is plotted as a function of elevation from a specified azimuth, is also of interest and may be similarly specified. *Synonym* main beam.

main station. A user instrument, *e.g.*, telephone set or terminal, with a distinct call number designation, connected to a local loop, used for originating calls, and on which incoming calls from the exchange are answered.

main storage. In a computer, program-addressable storage from which instructions and other data may be loaded directly into registers for subsequent execution or processing. *Note 1:* Main storage includes the total program-addressable execution space that may include one or more storage devices. *Note 2: "Main storage"* usually refers to large and intermediate computers, whereas *"memory"* usually refers to microcomputers, minicomputers, and calculators.

maintainability. 1. A characteristic of design and installation, expressed as the probability that an item will be retained in or restored to a specified condition within a given period of time, when the maintenance is performed in accordance with prescribed procedures and resources. 2. The ease with which maintenance of a functional unit can be performed in accordance with prescribed requirements.

maintenance. 1. Any activity, such as tests, measurements, replacements, adjustments and repairs, intended to restore or retain a functional unit in a specified state in which the unit can perform its required functions. 2. [For materiel], All action taken to retain materiel in a serviceable condition or to restore it to serviceability. It includes inspection, testing, servicing, classification as to serviceability, repair, rebuilding, and reclamation. 3. [For materiel], All supply and repair action taken to keep a force in condition to carry out its mission. 4. [For materiel], The routine recurring work required to keep a facility (plant, building, structure, ground facility, utility system, or other real property) in such condition that it may be continuously used, at its original or designed capacity and efficiency for its intended purpose.

maintenance control circuit (MCC). In a communications link, a circuit used by maintenance personnel for coordination. *Note:* An MCC is not available to operations or technical control personnel.

major lobe. *See* main lobe.

MAN. *Acronym for* metropolitan area network.

managed object. 1. In a network, an abstract representation of network resources that are managed. *Note:* A managed object may represent a physical entity, a network service, or an abstraction of a resource that exists independently of its use in management. 2. In telecommunications management, a resource within the telecommunications environment that may be managed through the use of operation, administration, maintenance, and provisioning application protocols.

management information system (MIS). An organized assembly of resources and procedures required to collect, process, and distribute data for use in decision making.

Manchester code. A code in which (a) data and clock signals are combined to form a single self-synchronizing data stream, (b) each encoded bit contains a transition at the midpoint of a bit period, (c) the direction of transition determines whether the bit is a "0" or a "1," and (d) the first half is the true bit value and the second half is the complement of the true bit value. *Contrast with* **non-return-to-zero.**

mandrel wrapping. In multimode fiber optics, a technique used to modify the modal distribution of a propagating optical signal. *Note:* A cylindrical rod wrap consists of a specified number turns of fiber on a mandrel of specified size, depending on the fiber characteristics and the desired modal distribution. It has application in optical transmission performance tests, to simulate, *i.e.,* establish, equilibrium mode distribution in a launch fiber (a fiber used to inject a test signal in another fiber that is under test). If the launch fiber is fully filled ahead of the mandrel wrap, the higher-order modes will be stripped off, leaving only lower-order modes. If the launch fiber is underfilled, *e.g.,* as a consequence of being energized by a laser diode or edge-emitting LED, there will be a redistribution to higher-order modes until modal equilibrium is reached.

man-machine system. A system in which the functions of a human operator and a machine are integrated.

margin. 1. In communications systems, the maximum degree of signal distortion that can be tolerated without affecting the restitution, *i.e.,* without its being interpreted incorrectly by the decision circuit. 2. The allowable error rate, deviation from normal, or degradation of the performance of, a system or device.

marine broadcast station. A coast station which makes scheduled broadcasts of time, meteorological, and hydrographic information.

marine utility station. A station in the maritime mobile service consisting of one or more hand-held radiotelephone units licensed under a single authorization. Each unit is capable of operation while being hand-carried by an individual.

maritime air communications. Communications systems, procedures, operations, and equipment that are used for message traffic between aircraft stations and ship stations in the maritime service. *Note:* Commercial, private, naval, and other ships are included in maritime air communications.

maritime broadcast communications net. A communications net that is used for international distress calling, including international lifeboat, lifecraft, and survival-craft high-frequency (HF); aeronautical emergency very high-frequency (VHF); survival ultra high-frequency (UHF); international calling and safety very high-frequency (VHF); combined scene-of-search-and-rescue; and other similar and related purposes. *Note:* Basic international distress calling is performed at either medium frequency (MF) or at high frequency (HF).

maritime mobile-satellite service. A mobile-satellite service in which mobile Earth stations are located on board ships; survival craft stations and emergency position-indicating radiobeacon stations may also participate in this service.

maritime mobile service. A mobile service between coast stations and ship stations, or between ship stations, or between associated on-board communication stations; survival craft stations and emergency position-indicating radiobeacon stations may also participate in this service.

maritime radionavigation-satellite service. A radionavigation-satellite service in which Earth stations are located on board ships.

maritime radionavigation service. A radionavigation service intended for the benefit and for the safe operation of ships.

mark. 1. In telegraphy, one of the two significant conditions of encoding. *Note 1:* The complementary significant condition is called a *"space." Note 2:* In modern digital communications, the two corresponding significant conditions of encoding are called "1" and "0." *Synonyms* marking pulse, marking signal. 2. A symbol or symbols that indicate the beginning or the end of a field, of a word, or of a data item in a file, record, or block.

marker beacon. A transmitter in the aeronautical radionavigation service which radiates vertically a distinctive pattern for providing position information to aircraft.

marking bias. The uniform lengthening of all marking signal pulse widths at the expense of the pulse widths of all spacing pulses.

marking end distortion. *See* end distortion.

marking pulse. *Synonym* mark.

marking signal. *Synonym* mark.

***m*-ary code.** *See n*-ary code.

***m*-ary signaling.** *See n*-ary code.

maser. *Acronym for* microwave amplification by stimulated emission of radiation. A member of the general class of microwave oscillators based on molecular interaction with electromagnetic radiation.

mask. 1. In communications systems, to obscure, hide, or otherwise prevent information from being derived from a signal. *Note 1:* Masking is usually the result of interaction with another signal, such as noise, static, jamming, or other forms of interference. *Note 2:* Masking is not synonymous with erasing or deleting. 2. In computing and data processing systems, a pattern of bits that can be used to retain or suppress segments of another pattern of bits.

masked threshold. The level at which an indistinguishable signal of interest becomes distinguishable from other signals or noise. *Note:* In acoustics, the masked threshold is usually expressed in dB.

master clock. A device that generates periodic, accurately spaced signals that are used for such purposes as timing, regulation of the operations of a processor, or generation of interrupts.

master frequency generator. In frequency-division multiplexing (FDM), equipment used to provide system end-to-end carrier frequency synchronization and frequency accuracy of tones. *Note:* The following types of oscillators are used in the Defense Communications System FDM systems:

Type 1 - A master carrier oscillator as an integral part of the multiplexer set.

Type 2 - A submaster oscillator equipment or slave oscillator equipment as an integral part of the multiplexer set.

Type 3 - An external master oscillator equipment that has extremely accurate and stable characteristics.

mastergroup. *See* group.

master oscillator. *Synonym* master frequency generator.

master-slave timing. Timing in which one station or node supplies the timing reference for all other interconnected stations or nodes.

master station. 1. In a data network, the station that is designated by the control station to ensure data transfer to one or more slave stations. *Note:* A master station controls one or more data links of the data communications network at any given instant. The assignment of master status to a given station is temporary and is controlled by the control station according to the procedures set forth in the operational protocol. Master status is normally conferred upon a station so that it may transmit a message, but a station need not have a message to send to be designated the master station. 2. In navigation systems using precise time dissemination, a station that has the clock used to synchronize the clocks of subordinate stations. 3. In basic mode link control, the data station that has accepted an invitation to ensure a data transfer to one or more slave stations. *Note:* At a given instant, there can be only one master station on a data link.

matched junction. A waveguide component having four or more ports, and so arranged that if all ports except one are terminated in the correct impedance, there will be no reflection of energy from the junction when the fourth port is driven by a transmission line having a matching impedance.

matching gel. *See* gel.

material absorption. *See* absorption.

material dispersion. *See* dispersion.

material dispersion coefficient [$M(\lambda)$]. In an optical fiber, pulse broadening per unit length of fiber and per unit of spectral width, usually expressed in picoseconds per (nanometer•kilometer). *Note 1:* For many optical fiber materials, $M(\lambda)$ approaches zero at a specific wavelength λ_0 between 1.3 μm and 1.5 μm. At wavelengths shorter than λ_0, $M(\lambda)$ is negative and increases with wavelength; at wavelengths longer than

λ_0, $M(\lambda)$ is positive and decreases with wavelength. *Note 2:* Pulse broadening caused by material dispersion in a unit length of optical fiber is given by the product of $M(\lambda)$ and spectral width ($\Delta\lambda$).

$$M(\lambda) = \frac{1}{c}\frac{dN}{d\lambda} = -\frac{1}{c}\frac{d^2n}{d\lambda^2},$$

where n is the refractive index of the material, N is the group index expressed as

$$N = n - \lambda\frac{dn}{d\lambda},$$

λ is the wavelength of interest, and c is the velocity of light in vacuum.

material scattering. Of an electromagnetic wave, scattering that is attributable to the intrinsic properties of the material through which the wave is propagating. *Note 1:* Ionospheric scattering and Rayleigh scattering are examples of material scattering. *Note 2:* In an optical fiber, material scattering is caused by micro-inhomogeneities in the refractive indices of the materials used to fabricate the fiber, including the dopants used to modify the refractive index profile.

maximal-ratio combiner. A diversity combiner in which (a) the signals from each channel are added together, (b) the gain of each channel is made proportional to the rms signal level and inversely proportional to the mean square noise level in that channel, and (c) the same proportionality constant is used for all channels. *Synonyms* ratio-squared combiner, post-detection combiner, predetection combining, selective combiner.

maximum block transfer time. The maximum allowable waiting time between initiation of a block transfer attempt and completion of a successful block transfer.

maximum calling area. Geographic calling limits permitted to a particular access line based on requirements for the particular line. *Note:* Maximum calling area restrictions are imposed for network control purposes.

maximum disengagement time. The maximum allowable waiting time between initiation of a disengagement attempt and successful disengagement.

maximum justification rate. *Synonym* maximum stuffing rate.

maximum keying frequency. In facsimile systems, the frequency in hertz numerically equal to the spot speed divided by twice the X-dimension of the scanning spot.

maximum modulating frequency. In a facsimile transmission system, the highest picture frequency that is required. *Note:* The maximum modulating frequency and the maximum keying frequency are not necessarily equal.

maximum stuffing rate. In a bit stream, the maximum rate at which stuffing bits can be inserted into the stream. *Synonym* maximum justification rate.

maximum usable frequency (MUF). In radio transmission using reflection from the regular ionized layers of the ionosphere, the upper frequency limit that can be used for transmission between two points at a specified time. *Note:* MUF is a median frequency applicable to 50% of the days of a month, as opposed to 90% cited for the lowest usable high frequency (LUF) and the optimum traffic frequency (FOT).

maximum user signaling rate. The maximum rate, in bits per second, at which binary information can be transferred in a given direction between users over the telecommunications system facilities dedicated to a particular information transfer transaction, under conditions of continuous transmission and no overhead information. *Note 1:* For a single channel, the signaling rate is given by

$$SCSR = \frac{\log_2 n}{T},$$

where *SCSR* is the single-channel signaling rate in bits per second, *T* is the minimum time interval in seconds for which each level must be maintained, and n is the number of significant conditions of modulation of the channel. *Note 2:* In the case where an individual end-to-end telecommunications service is provided by parallel channels, the parallel-channel signaling rate is given by

$$PCSR = \sum_{i-1}^{m} \frac{\log_2 n_i}{T_i},$$

where *PCSR* is the total signaling rate for *m* channels, *m* is the number of parallel channels, T_i is the minimum interval between significant instants for the *I*-th channel, and n_i is the number of significant conditions of modulation for the *I*-th channel. *Note 3:* In the case where an end-to-end telecommunications service is provided by tandem channels, the end-to-end signaling rate is the lowest signaling rate among the component channels.

Maxwell's equations. A set of partial differential equations that describe and predict the behavior of electromagnetic waves in free space, in dielectrics, and at conductor-dielectric boundaries. *Note:* Maxwell's equations expand upon and unify the laws of Ampere, Faraday, and Gauss, and form the foundation of modern electromagnetic theory.

MCC. *Abbreviation for* maintenance control circuit.

MCM. *Abbreviation for* multicarrier modulation.

MDF. *Abbreviation for* main distribution frame.

meaconing. A system of receiving radio beacon signals and rebroadcasting them on the same frequency to confuse navigation. The meaconing stations cause inaccurate bearings to be obtained by aircraft or ground stations.

mean power (of a radio transmitter). The average power supplied to the antenna transmission line by a transmitter during an interval of time sufficiently long compared with the lowest frequency encountered in the modulation taken under normal operating conditions. *Note:* Normally, a time of 0.1 second, during which the mean power is greatest, will be selected.

mean time between failures (MTBF). 1. An indicator of expected system reliability calculated on a statistical basis from the known failure rates of various components of the system. *Note:* .MTBF is usually expressed in hours. 2. Of a system, over a long performance measurement period, the measurement period divided by the number of failures that have occurred during the measurement period. 3. For population of items, during a measurement period, the total functioning life of the population of items divided by the total number of failures within the population during the measurement period. *Note 1:* The total functioning life of the population may be calculated as the summation of the operating life of every item in the population over the measurement period. When computing the MTBF, any measure of operating life may be used, such as time, cycles, kilometers, or events. *Note 2:* For example, if a total of 1,000 events, such as data transfers, radio transmissions, or system boots, occurs in a population of items during a measurement period of 100 hours and there are a total of 10 failures among the entire population, the MTBF for each item is (1000)(100)/ 10 = 10^4 hours.

mean time between outages (*MTBO*). In a system, the mean time between equipment failures that result in loss of system continuity or unacceptable degradation. *Note:* The *MTBO* is calculated by the equation,

$$\text{MTBO} = \frac{\text{MTBF}}{1 - \text{FFAS}},$$

where *MTBF* is the nonredundant mean time between failures and *FFAS* is the fraction of failures for which the failed equipment is automatically bypassed.

mean time to repair (MTTR). The total corrective maintenance time divided by the total number of corrective maintenance actions during a given period of time.

mean time to service restoral (MTSR). The mean time to restore service following system failures that result in a service outage. *Note:* The time to restore includes all time from the occurrence of the failure until the restoral of service.

measured-rate service. Telephone service for which charges are made in accordance with the total connection time of the line.

measurement period. *See* performance measurement period.

mechanically induced modulation. Optical signal modulation induced by mechanical means. *Note:* An example of deleterious mechanically induced modulation is speckle noise created in a multimode fiber by an imperfect splice or imperfectly mated connectors. Mechanical disturbance of the fiber ahead of the joint will introduce changes in the modal structure, resulting in variations of joint loss.

mechanically intermateable connectors. Connectors that are mechanically mateable, without creating mechanical damage, and without regard to attenuation properties.

mechanical splice. Of optical fibers, a splice, *i.e.*, permanent joint, accomplished by aligning the mating fibers in some kind of mechanical fixture. *Note 1:* The fibers may be secured by mechanical means or with an optical adhesive. *Note 2:* When the fibers are secured by mechanical means, the gap between them is usually filled with an index-matching gel to reduce Fresnel reflection. Likewise, the optical adhesives that are used in conjunction with mechanical splices are formulated to have a refractive index that approximates that of the glass, and also serve to reduce Fresnel reflection.

mediation function. In telecommunications network management, a function that routes or acts on information passing between network elements and network operations. *Note 1:* Examples of mediation functions are communications control, protocol conversion, data handling, communications of primitives, processing that includes decision-making, and data storage. *Note 2:* Mediation functions can be shared among network elements, mediation devices, and network operation centers.

medium. 1. In telecommunications, the transmission path along which a signal propagates, such as a wire pair, coaxial cable, waveguide, optical fiber, or radio path. 2. The material on which data are or may be recorded, such as plain paper, paper tapes, punched cards, magnetic tapes, magnetic disks, or optical disks.

medium access control (MAC) sublayer. In a communications network, the part of the data link layer that supports topology-dependent functions and uses the services of the physical layer to provide services to the logical link control sublayer.

medium access unit (MAU). In a communications system, the equipment that adapts or formats the signal for transmittal over the communication medium. *Note:* An example of a MAU is an optical transmitter, which accepts an electrical signal at its input port and converts it to an optical signal accessible at its output port.

medium frequency (MF). Frequencies from 300 kHz to 3000 kHz.

medium interface connector (MIC). In communications systems, the connector at the interface point between the bus interface unit and the terminal, *i.e.*, the medium interface point.

medium interface point (MIP). In communication systems, the location at which the standards for the interface parameters between a terminal and the line facility are implemented.

medium-power talker. A hypothetical talker, within a log-normal distribution of talkers, whose volume lies at the medium power of all talkers determining the volume distribution at the point of interest. *Note:* When the distribution follows a log-normal curve (values expressed in decibels), the mean and standard deviation can be used to compute the medium-power talker. The talker volume distribution follows a log-normal curve and the medium-power talker is uniquely determined by the average talker volume. The medium-power talker volume, V, is given by $V = V_0 + 0.115\sigma^2$, where V_0 is the average of the talker volume distribution in volume units (vu), and σ^2 is the variance of the distribution.

megahertz (MHz). A unit of frequency denoting one million (10^6) Hz.

memory. 1. All of the addressable storage space in a processing unit and other internal memory that is used to execute instructions. 2. *Loosely*, the volatile, main storage in computers. *See* **random access memory.** *Contrast with* hard disk.

menu. A displayed list of options from which a user selects actions to be performed.

MERCAST. *Acronym for* merchant-ship broadcast system.

merchant-ship broadcast system (MERCAST). A maritime shore-to-ship broadcast system in which the ocean areas are divided into primary broadcast areas each covered by a high-powered shore radio station that broadcasts simultaneously on one medium frequency (MF) and one or more high frequencies (HF) for routing messages to ocean-going ships. *Note:* In some instances, coast stations may repeat the messages.

meridional ray. In fiber optics, a ray that passes through the optical axis of an optical fiber (in contrast with a skew ray, which does not).

mesh network. *See* network topology.

mesh topology. *See* network topology.

mesochronous. The relationship between two signals such that their corresponding significant instants occur at the same average rate.

message. 1. Any thought or idea expressed briefly in a plain or secret language, prepared in a form suitable for transmission by any means of communication. *Note:*

A message may be a one-unit message or a multiunit message. 2. [In telecommunications,] Record information expressed in plain or encrypted language and prepared in a format specified for intended transmission by a telecommunications system. 3. An arbitrary amount of information whose beginning and end are defined or implied.

message alignment indicator. In a signal message, data transmitted between the user part and the message transfer part to identify the boundaries of the signal message.

message broadcast. An electronic-mail conference capability using data terminals. *Note:* Control can be maintained by the user or by the network.

message center. *See* communications center.

message feedback. A method of checking the accuracy of transmission of data by sending received data back to the sending end for comparison with the original data that have been stored there for this purpose.

message format. A predetermined or prescribed spatial or time-sequential arrangement of the parts of a message that is recorded in or on a data storage medium. *Note:* Messages prepared for electrical transmission are usually composed on a printed blank form with spaces for each part of the message and for administrative entries.

message handling system (MHS). In the CCITT X.400 Recommendations, the family of services and protocols that provides the functions for global electronic-mail transfer among local mail systems.

message heading. In radio communications, the message part or parts that (a) precede the text, *i.e.*, the message body, in time or space according to established conventions and (b) may include several data items, such as address groups, routing indicators, action addressee designators, information addressee designators, exempted addressee designators, prosigns, prowords, clear indicators, date-time groups, originator designators, special instructions, and protocol symbols. *Note:* Several message heading data items may be combined into a message preamble.

message part. 1. In radio communications, one of the three major subdivisions of a message, namely the heading, the text, or the ending. *Note:* Each message part may have separate components and each component may have elements and contents. 2. In cryptosystems, text that results from the division of a long message into several shorter messages of different lengths as a transmission security measure. *Note:* Message parts are usually prepared in such a manner as to appear unrelated externally. Statements that identify the parts for assembly at reception are encrypted in the texts.

message register leads. Terminal equipment leads at the interface used solely for receiving dc message register pulses from a central office at a PBX so that message unit information normally recorded at the central office only is also recorded at the PBX. . . .

message service. Switched service furnished to the general public (as distinguished from private line service). Except as otherwise provided, this includes exchange switched services and all switched services provided by interexchange carriers and completed by a local telephone company's access services. . . . *Synonym* message toll service.

message switching. A method of handling message traffic through a switching center, either from local users or from other switching centers, whereby the message traffic is stored and forwarded through the system.

message toll service. *Synonym* message service.

message transfer part. The part of a common-channel signaling system that transfers signal messages and performs associated functions, such as error control and signaling link security.

message unit. A unit of measure for charging telephone calls, based on parameters such as the length of the call, the distance called, and/or the time of day.

messaging service. In integrated services digital networks (ISDN), an interactive telecommunications service that provides for information interchange among users by means of store-and-forward, electronic mail, or message-handling functions.

metallic circuit. A circuit in which metallic conductors are used and in which the ground or earth forms no part.

metallic voltage. A potential difference between metallic conductors, as opposed to a potential difference between a metallic conductor and ground.

meteor burst communications. Communications by the propagation of radio signals reflected by ionized meteor trails.

meteorological aids service. A radiocommunication service used for meteorological, including hydrological, observations and exploration.

meteorological-satellite service. An Earth exploration-satellite service for meteorological purposes.

metric system. A decimal system of weights and measures based on the meter as a unit of length and the kilogram as a unit of mass. *Note:* The modern form of the metric system is the International System of Units (SI). *See* International System of Units.

metropolitan area network (MAN). A data communications network that (a) covers an area larger than a campus area network and smaller than a wide area network (WAN), (b) interconnects two or more LANs, and (c) usually covers an entire metropolitan area, such as a large city and its suburbs.

MF. *Abbreviation for* medium frequency.

MFD. *Abbreviation for* mode field diameter.

MFJ. *Abbreviation for* Modification of Final Judgment.

MFSK. *Abbreviation for* multiple frequency-shift keying.

MHS. *Abbreviation for* message handling system.

MHz. *Abbreviation for* megahertz.

MIC. *Abbreviation for* medium interface connector.

microbend. In an optical waveguide, sharp curvatures involving local axial displacements of a few micrometers and spatial wavelengths of a few millimeters. *Note:* Microbends can result from waveguide coating, cabling, packaging, and installation. Microbending can cause significant radiative loss and mode coupling.

microbending. *See* microbend.

microbend loss. In an optical fiber, the optical power loss caused by a microbend.

microcircuit. *Synonym* integrated circuit.

microcode. A sequence of microinstructions that is fixed in storage that is not program-addressable, and that performs specific processing functions.

microcomputer. A computer (a) in which the processing unit is a microprocessor and (b) that usually consists of a microprocessor, a storage unit, an input channel, and an output channel, all of which may be on one chip.

microfinishing film. A film of dimensionally stable plastic, to which are adhered carefully graded abrasive or polishing powders, *i.e.*, particles, having dimensions in the micrometer or submicrometer range. *Note:* Microfinishing films resemble sandpaper, but have much smaller abrasive or polishing particles. They are used commercially to shape and/or polish machined parts. They are also used to finish the endfaces of certain types of optical connectors.

microinstruction. An instruction that controls data flow and instruction-execution sequencing in a processor at a more fundamental level than machine instructions. *Note:* A series of microinstructions is necessary to perform an individual machine instruction.

micro-mainframe link. A physical or logical connection established between a remote microprocessor and mainframe host computer for the express purpose of uploading, downloading, or viewing interactive data and databases on-line in real time. *Note:* A micro-mainframe link usually requires terminal emulation software on the microcomputer.

microprocessor. A central processing unit implemented on a single chip.

microprogram. A sequence of microinstructions that are in special storage where they can be dynamically accessed to perform various functions.

microwave (mw). Loosely, an electromagnetic wave having a wavelength from 300 mm to 10 mm (1 GHz to 30 GHz). *Note:* Microwaves exhibit many of the properties usually associated with waves in the optical regime, *e.g.*, they are easily concentrated into a beam.

Mie scattering. Scattering of an electromagnetic wave by particles or refractive index inhomogeneities of a size on the order of the wavelength of interest.

mileage. In telecommunications, a specified distance used in tariff calculations, *i.e.*, toll charge calculations. *Note:* Mileage is locally defined and often refers to airline distance rather than actual communication system route miles.

military common emergency frequency. A frequency that (a) is used by all military units that are equipped to operate at that frequency or in the band in which that frequency lies and (b) is also used internationally by survival-craft stations and survival-craft equipment.

millimeter wave. Loosely, an electromagnetic wave having a wavelength from 1 mm to 0.1 mm (300 GHz to 3000 GHz). *Note:* Millimeter waves exhibit many of the properties usually associated with waves in the optical regime, *e.g.*, they are easily concentrated into a beam.

minicomputer. *See* computer.

minimize. A condition wherein normal message and telephone traffic is drastically reduced in order that messages connected with an actual or simulated emergency shall not be delayed.

minimum bend radius. The radius below which an optical fiber or fiber-optic cable should not be bent. *Note 1:* The minimum bend radius is of particular importance in the handling of fiber-optic cables. It will vary with different cable designs. The manufacturer should specify the minimum radius to which the cable may safely be bent during installation, and for the long term. The former is somewhat shorter than the latter. *Note 2:* The minimum bend radius is in general also a function of tensile stresses, *e.g.*, during installation, while being bent around a sheave while the fiber or cable is under tension. *Note 3:* If no minimum bend radius is specified, one is usually safe in assuming a minimum long-term low-stress radius not less than 15 times the cable diameter.

minimum discernable signal (MDS). *See* threshold.

minimum-dispersion slope. *See* zero-dispersion slope.

minimum-dispersion wavelength. *Synonym* zero-dispersion wavelength.

minimum-dispersion window. 1. The window of an optical fiber at which material dispersion is very small. *Note 1:* In silica-based fibers, the minimum-dispersion window occurs at a wavelength of approximately 1.3 μm. *Note 2:* The minimum-dispersion window may be shifted toward the minimum-loss window, *i.e.*, 1.55 μm, by the addition of dopants during manufacture. 2. In a single-mode fiber, the window or, in the case of doubly or quadruply clad fibers, windows, at which material and waveguide dispersion cancel one another, resulting in extremely wide bandwidth, *i.e.*, extremely low dispersion, over a very narrow range of wavelengths. *Synonym* zero dispersion window.

minimum-loss window. Of an optical fiber, the transmission window at which the attenuation coefficient is at or near the theoretical (quantum-limited) minimum. *Note 1:* If the losses from various mechanisms are plotted on a single graph as a function of wavelength, the minimum-loss window occurs in the vicinity of the wavelength at which the Rayleigh-scattering attenuation curve and the infrared-phonon-absorption curve intersect. *Note 2:* For silica-based fibers, the minimum-loss window occurs at approximately 1.55 μm.

minimum picture interval. The minimum time between the television pictures that have been selected for encoding. *Note:* CCITT Recommendation H.221 cites the following values for picture interval: 1/29.97, 2/29.97, 3/29.97, and 4/29.97 seconds per picture.

MIP. *Abbreviation for* medium interface point.

MIS. *Abbreviation for* management information system.

misalignment loss. *See* angular misalignment loss, gap loss, lateral offset loss.

misdelivered block. A block received by a user other than the one intended by the message source.

mission bit stream. *Synonym* payload.

mixer. A nonlinear circuit or device that accepts as its input two different frequencies and presents at its output (a) a signal equal in frequency to the sum of the frequencies of the input signals, (b) a signal equal in frequency to the difference between the frequencies of the input signals, and, if they are not filtered out, (c) the original input frequencies.

mixing. *See* heterodye.

MLPP. *Abbreviation for* multilevel precedence and preemption.

MM patch bay. A patching facility designed for patching and monitoring of digital data circuits at rates exceeding 3 Mb/s.

mobile Earth station. An Earth station in the mobile-satellite service intended to be used while in motion or during halts at unspecified points.

mobile-satellite service. A radiocommunication service:

- between mobile earth stations and one or more space stations, or between space stations used by this service; or

- between mobile earth stations by means of one or more space stations. This service may also include feeder links necessary for its operation.

mobile service. A radiocommunication service between mobile and land stations, or between mobile stations.

mobile services switching center (MSC). In an automatic cellular mobile system, the interface between the radio system and the public switched telephone network. *Note:* The MSC performs all signaling functions that are necessary to establish calls to and from mobile stations.

mobile station. A station in the mobile service intended to be used while in motion or during halts at unspecified points.

modal dispersion. *Incorrect synonym for* multimode distortion.

modal distortion. *Synonym* **mu**ltimode distortion.

modal distribution. In an optical waveguide operating at a given wavelength, the number of modes supported, and their propagation time differences.

modal loss. In an open waveguide, such as an optical fiber, a loss of energy on the part of an electromagnetic wave due to obstacles outside the waveguide, abrupt changes in direction of the waveguide, or other anomalies, that cause changes in the propagation mode of the wave in the waveguide.

modal noise. Noise generated in an optical fiber system by the combination of mode-dependent optical losses and fluctuation in the distribution of optical energy among the guided modes or in the relative phases of the guided modes. *Synonym* speckle noise.

mode. 1. In a waveguide or cavity, one of the various possible patterns of propagating or standing electromagnetic fields. *Note 1:* Each mode is characterized by frequency, polarization, electric field strength, and magnetic field strength. *Note 2:* The electromagnetic field pattern of a mode depends on the frequency, refractive indices or dielectric constants, and waveguide or cavity geometry. 2. Any electromagnetic field distribution that satisfies Maxwell's equations and the applicable boundary conditions. 3. In data communications, a protocol used to transfer data from switch to switch or from switch to terminal. 4. In statistics, the value associated with the highest peak in a probability density function.

mode coupling. In an electromagnetic waveguide, the exchange of power among modes. *Note:* In a multimode optical fiber, mode coupling reaches statistical equilibrium, *i.e.*, equilibrium mode distribution, after the equilibrium length has been traversed.

mode field diameter (MFD). An expression of distribution of the irradiance, *i.e.*, the optical power, across the end face of a single-mode fiber. *Note:* For a Gaussian power distribution in a single-mode optical fiber, the mode field diameter is that at which the electric and magnetic field strengths are reduced to 1/ e of their maximum values, *i.e.,* the diameter at which power is reduced to $1/e^2$ of the maximum power, because the power is proportional to the square of the field strength.

mode filter. A device used to select, reject, or attenuate a certain mode or modes.

mode [identification friend or foe]. The number or letter referring to the specific pulse spacing of the signals transmitted by an interrogator.

modem. *Acronym for* modulator/demodulator. 1. In general, a device that both modulates and demodulates signals. 2. In computer communications, a device used for converting digital, signals into, and recovering them from, quasi-analog signals suitable for transmission over analog communications channels. *Note:* Many additional functions may be added to a modem to provide for customer service and control features. *Synonym* signal conversion equipment. 3. In FDM carrier systems, a device that converts the voice band to, and recovers it from, the first level of frequency translation.

mode mixer. *Synonym* mode scrambler.

modem patch. A method of electrically interconnecting circuits by using back-to-back modems.

mode partition noise. In an optical communications link, phase jitter of the signal caused by the combined effects of mode hopping in the optical source and intramodal distortion in the fiber. *Note:* Mode hopping causes random wavelength changes which in turn affect the group velocity, *i.e.*, the propagation time. Over a long length of fiber, the cumulative effect is to create jitter, *i.e.*, mode partition noise. The variation of group velocity creates the mode partition noise.

mode scrambler. 1. A device for inducing mode coupling in an optical fiber. 2. A device composed of one or more optical fibers in which strong mode coupling occurs. *Note:* Mode scramblers are used to provide a modal distribution that is independent of the optical source, for purposes of laboratory or field measurements or tests. *Synonym* mode mixer.

mode stripper. *See* cladding mode stripper.

mode volume. The number of bound modes that an optical fiber is capable of supporting. *Note:* The mode volume M is approximately given by $V^2/2$ and $(V^2/2)[g/(g+2)]$ respectively for step-index and power-law profile fibers, where g is the profile parameter, and V is the normalized frequency greater than 5.

Modification of Final Judgment (MFJ). The 1982 antitrust suit settlement agreement (*"Consent Decree"*) entered into by the United States Department of Justice and the American Telephone and Telegraph Company (AT&T) that, after modification and upon approval of the United States District Court for the District of Columbia, required the divestiture of the Bell Operating Companies from AT&T.

modified AMI code. *Abbreviation for* modified alternate mark inversion code. A T-carrier AMI line code in which bipolar violations may be deliberately inserted to maintain system synchronization. *Note 1:* The clock rate of an incoming T-carrier signal is extracted from its bipolar line code. T-carrier was originally developed for voice applications. When voice signals are digitized for transmission via T-carrier, there is no problem in maintaining system synchronization, because of the nature of the digitized signals. However, when used for the transmission of digital data, the conventional AMI line code may fail to have sufficient marks, *i.e.*, "1's," to permit recovery of the incoming clock, and synchronization is lost. This happens when there are too many consecutive zeros in the user data being transported. To prevent loss of synchronization when a long string of zeros is present in the user data, deliberate bipolar violations are inserted into the line code, to create a sufficient number of marks to maintain synchronization. The receive terminal equipment recognizes the bipolar violations and removes from the user data the marks attributable to the bipolar violations. *Note 2:* The exact pattern of bipolar violations that is transmitted in any given case depends on the line rate and the polarity of the last valid mark in the user data prior to the unacceptably long string of zeros. *Note 3:* The number of consecutive zeros that can be tolerated in user data depends on the data rate, *i.e.*, the level of the line code in the T-carrier hierarchy. The North American T1 line code (1.544 Mb/s) does not use bipolar violations. The European T1 line code (2.048 Mb/s) may use bipolar violations when 8 or more consecutive zeros are present. This line code is called *bipolar with eight-zero substitution (B8ZS)*. (In all levels of the European T-carrier hierarchy, the patterns of bipolar violations that are used differ from those used in the North American hierarchy.) At the North American T2 rate (6.312 Mb/s), bipolar violations are inserted if 6 or more consecutive zeros occur. This line code is called *bipolar with six-zero substitution (B6ZS).* At the North American T3 rate (44.736 Mb/s), bipolar violations are inserted if 3 or more consecutive zeros occur. This line code is called *"bipolar with three-zero substitution" (B3ZS).*

modular. Pertaining to the design concept in which interchangeable units are used to create a functional end product.

modular jack. A device that conforms to the *Code of Federal Regulations,* Title 47, part 68, which defines the size and configuration of all units that are permitted for connection to the public exchange facilities.

modulation. The process, or result of the process, of varying a characteristic of a carrier, in accordance with an information-bearing signal.

modulation factor. In amplitude modulation, the ratio of the peak variation actually used, to the maximum design variation in a given type of modulation. *Note:* In conventional amplitude modulation, the maximum design variation is considered that for which the instantaneous amplitude of the modulated signal reaches zero. When zero is reached, the modulation is considered 100%.

modulation index. In angle modulation, the ratio of the frequency deviation of the modulated signal to the frequency of a sinusoidal modulating signal. *Note:* The modulation index is numerically equal to the phase deviation in radians.

modulation rate. 1. The rate at which a carrier is varied to represent the information in a digital signal. *Note:* Modulation rate and information transfer rate are not necessarily the same. 2. For modulated digital signals, the reciprocal of the unit interval of the modulated signal, measured in seconds.

modulation suppression. In the reception of an amplitude-modulated signal, an apparent reduction in the depth of modulation of a wanted signal, caused by the presence, at the detector, of a stronger unwanted signal.

modulator. A device that imposes a signal on a carrier.

modulator-demodulator (modem). *See* modem.

module. 1. An interchangeable subassembly that constitutes part of, *i.e.,* is integrated into, a larger device or system. 2. In computer programming, a program unit that is discrete and identifiable with respect to compiling, combining with other modules, and loading.

monitor. 1. Software or hardware that is used to scrutinize and to display, record, supervise, control, or verify the operations of a system. *Note:* Possible uses of monitors are to indicate significant departures from the norm, or to determine levels of utilization of particular functional units. 2. A device used for the real-time temporary display of computer output data. *Note:* Monitors usually use cathode-ray-tube or liquid-crystal technology. *Synonyms* video display terminal, video display unit, visual display unit.

monitoring. 1. The act of listening, carrying out surveillance on, and/or recording the emissions of one's own or allied forces for the purpose of maintaining and improving procedural standards and security, or for reference, as applicable. 2. The act of listening, carrying out surveillance on, and/or recording of enemy emissions for intelligence purposes. 3. The act of detecting the presence of signals, such as electromagnetic radiation, sound, or visual signals, and the measurement thereof with appropriate measuring instruments. 4. The act of detecting the presence of radiation and the measurement thereof with radiation measuring instruments. *Synonym* radiological monitoring.

monitor jack. A jack used to access communications circuits to observe signal conditions without interrupting the services.

monitor key. A key used to access communications circuits to observe signal conditions without interrupting the services.

monochromatic. In optics, pertaining to a single wavelength of electromagnetic radiation or to a single color. *Note:* In practice, optical radiation is never perfectly monochromatic, *i.e.*, it never consists of only one wavelength. It always has a finite spectral width, albeit narrow.

monochromator. In optics, an instrument for isolating narrow portions of the spectrum.

monomode optical fiber. *Synonym* single-mode optical fiber.

Mosaic. A portable World Wide Web browser that provides a graphical user interface to hypertext-based information.

mosquito noise. In a video display, distortion sometimes seen around the edges of moving objects, and characterized by moving artifacts around edges and/or by blotchy noise patterns superimposed over the objects, resembling a mosquito flying around a person's head and shoulders.

motion compensation. Interframe coding that (a) is used to compress motion of video images and (b) uses an algorithm to examine a sequence of image frames to measure the difference from frame to frame in order to send motion vector information.

motion response degradation. The deterioration of motion video quality, resulting in a loss of perceived spatiotemporal resolution.

motion video. In video systems, temporally varying visual imagery intended to communicate or to convey movement or change.

mouse. A hand-held computer input device that generates signals that increment, *i.e.,* slew, the position of a cursor on a video display. *Note:* A mouse is placed on a flat surface and moved manually in the direction in which it is desired to move the cursor. A mouse has momentary switches ("buttons") that may be finger-operated to trigger an event after the cursor is positioned correctly.

M-patch bay. A patching facility designed for patching and monitoring digital data circuits at data signaling rates from 1 Mb/s (megabits per second) to 3 Mb/s.

m-sequence. *See n*-sequence.

MTBF. *Abbreviation for* mean time between failures.

MTBO. *Abbreviation for* mean time between outages.

MTSR. *Abbreviation for* mean time to service restoral.

MTTR. *Abbreviation for* mean time to repair.

mudbox. Equipment that is sufficiently rugged to withstand adverse environments. *Note:* A mudbox is expected to operate unsheltered on the ground.

MUF. *Abbreviation for* maximum usable frequency.

mu-law (μ-law). *See* mu-law (μ-law) algorithm.

mu-law (μ-law) algorithm. A standard analog signal compression algorithm, used in digital communications systems of the North American digital hierarchy, to optimize, *i.e.*, modify, the dynamic range of an analog signal prior to digitizing. *Note:* The wide dynamic range of speech does not lend itself well to efficient linear digital encoding. Mu-law encoding effectively reduces the dynamic range of the signal, thereby increasing the coding efficiency and resulting in a signal-to-distortion ratio that is greater than that obtained by linear encoding for a given number of bits.

muldem. *Acronym for* multiplexer/ demultiplexer.

multiaddress calling. A service feature that permits a user to designate more than one addressee for the same data. *Note:* Multiaddress calling may be performed sequentially or simultaneously.

multicarrier modulation (MCM). A technique of transmitting data by dividing the data into several interleaved bit streams and using these to modulate several carriers. *Note:* MCM is a form of frequency-division multiplexing.

multicast. 1. In a network, a technique that allows data, including packet form, to be simultaneously transmitted to a selected set of destinations. *Note:* Some networks, such as Ethernet, support multicast by allowing a network interface to belong to one or more multicast groups. 2. To transmit identical data simultaneously to a selected set of destinations in a network, usually without obtaining acknowledgement of receipt of the transmission.

multicast address. A routing address that (a) is used to address simultaneously all the computers in a group and (b) usually identifies a group of computers that share a common protocol, as opposed to a group of computers that share a common network. *Note:* Multicast address also applies to radio communications. *Synonym (in Internet protocol)* class d address.

multichannel. Pertaining to communications, usually full-duplex communications, on more than one channel. *Note:* Multichannel transmission may be accomplished by time-division multiplexing, frequency-division multiplexing, phase-division multiplexing, or space diversity.

multicoupler. In radio communications, a device for connecting several receivers or transmitters to one antenna in such a way that the equipment impedances are properly matched to the antenna impedance.

multi-element dipole antenna. An antenna consisting of an arrangement of multiple dipole antennas. *Note:* Various directivity patterns may be obtained by varying the arrangement of the dipoles and the way they are driven.

multifiber cable. A fiber-optic cable having two or more fibers, each of which is capable of serving as an independent optical transmission channel.

multifiber cable assembly. *See* cable assembly.

multifiber joint. An optical splice or connector designed to mate two multifiber cables, providing simultaneous optical alignment of all individual optical fibers.

multiframe. In PCM systems, a set of consecutive frames in which the position of each frame can be identified by reference to a multiframe alignment signal. *Note:* The multiframe alignment signal does not necessarily occur, in whole or in part, in each multiframe.

multifrequency pulsing. *Synonym* multifrequency signaling. *See* dual-tone multifrequency signaling.

multifrequency signaling. *Synonym* dual-tone multifrequency signaling.

multilayer filter. *See* interference filter.

multilevel modulation. *See* *n*-ary code.

multilevel precedence and preemption (MLPP). In military communications, a priority scheme (a) for assigning one of several precedence levels to specific calls or messages so that the system handles them in a predetermined order and time frame, (b) for gaining controlled access to network resources in which calls and messages can be preempted only by higher priority calls and messages, (c) that is recognized only within a predefined domain, and (d) in which the precedence level of a call outside the predefined domain is usually not recognized.

multilink operation. In packet-switched networks, the simultaneous use of multiple links for the transmission of different segments of the same message unit. *Note:* Use of multilink operation is intended to increase the effective rate of message transmission. Multilink operation requires special procedures for multiplexing/demultiplexing control.

multimedia. Pertaining to the processing and integrated presentation of information in more than one form, *e.g.*, video, voice, music, or data. *Contrast with* multiple media.

multimode dispersion. *Incorrect synonym for* multimode distortion.

multimode distortion. A distortion mechanism, occurring in multimode fibers, in which the signal is spread in time because the velocity of propagation of the optical signal is not the same for all modes. *Note 1:* In the ray-optics analogy, multimode distortion in a step-index optical fiber may be compared to multipath propagation of a radio signal. The direct signal is distorted by the arrival of the reflected signal a short time later. In a step-index optical fiber, rays taking more direct paths through the fiber core, *i.e.*, those which undergo the fewest reflections at the core-cladding boundary, will traverse the length of the fiber sooner than those rays which undergo more reflections. This results in distortion of the signal. *Note 2:* Multimode distortion limits the bandwidth of multimode fibers. For example, a typical step-index fiber with a 50-μm core would be limited to approximately 20 MHz for a one-kilometer length, *i.e.*, a bandwidth of 20 MHz•km. *Note 3:* Multimode distortion may be considerably reduced, but never completely eliminated, by the use of a core having a graded refractive index. The bandwidth of a typical off-the-shelf graded-index multimode fiber, having a 50-μm core, may approach 1 GHz•km or more. Multimode graded-index fibers having bandwidths approaching 3 GHz•km have been produced. *Note 4:* Because of its similarity to dispersion in its effect on the optical signal, multimode distortion is sometimes incorrectly referred to as *"intermodal dispersion," "modal dispersion,"* or *"multimode dispersion."* Such usage is incorrect because multimode distortion is not a truly dispersive effect. Dispersion is a wavelength-dependant phenomenon, whereas multimode distortion may occur at a single wavelength. *Synonyms* intermodal delay distortion, intermodal distortion, modal distortion.

multimode group delay. *Synonym* differential mode delay.

multimode optical fiber. An optical fiber that supports the propagation of more than one bound mode. *Note:* A multimode optical fiber may be either a graded-index (GI) fiber or a step-index (SI) fiber.

multiparty line. *Synonym* party line.

multipath. The propagation phenomenon that results in radio signals' reaching the receiving antenna by two or more paths. *Note 1:* Causes of multipath include atmospheric ducting, ionospheric reflection and refraction, and reflection from terrestrial objects, such as mountains and buildings. *Note 2:* The effects of multipath include constructive and destructive interference, and phase shifting of the signal. *Note 3:* In facsimile and television transmission, multipath causes jitter and ghosting.

multiple. A system of wiring so arranged that a circuit, a line, or a group of lines is accessible at a number of points. *Synonym* multipoint.

multiple access. 1. The connection of a user to two or more switching centers by separate access lines using a single message routing indicator or telephone number. 2. In satellite communications, the capability of a communications satellite to function as a portion of a communications link between more than one pair of satellite terminals concurrently. *Note:* The three types of multiple access presently used with communications satellites are code-division, frequency-division, and time-division multiple access. 3. In computer networking, a scheme that allows temporary access to the network by individual users, on a demand basis, for the purpose of transmitting information. *Note:* Examples of multiple access are carrier sense multiple access with collision avoidance (CSMA/CA) and carrier sense multiple access with collision detection (CSMA/CD).

multiple call. *Synonym* conference call.

multiple frequency-shift keying (MFSK). Frequency-shift keying (FSK) in which multiple codes are used in the transmission of digital signals. *Note:* In MFSK, the coding schemes use multiple frequencies that are transmitted concurrently or sequentially.

multiple homing. 1. In telephone systems, the connection of a terminal facility so that it can be served by one or several switching centers. *Note:* Multiple homing may use a single directory number. 2. In telephone systems, the connection of a terminal facility to more than one switching center by separate access lines. *Note:* Separate directory numbers are applicable to each switching center accessed.

multiple media. Transmission media using more than one type of transmission path (*e.g.*, optical fiber, radio, and copper wire) to deliver information. *Contrast with* multimedia.

multiple-spot scanning. In facsimile systems, scanning performed simultaneously by two or more scanning spots, each one analyzing its fraction of the total scanned area of the object.

multiplex (MUX). *See* multiplexing.

multiplex aggregate bit rate. In a time-division multiplexer, the bit rate that is equal to the sum of (a) the input channel data signaling rates available to the user and (b) the rate of the overhead bits required.

multiplex baseband. 1. In frequency-division multiplexing, the frequency band occupied by the aggregate of the signals in the line interconnecting the multiplexing and radio or line equipment. 2. In frequency division multiplexed carrier systems, at the input to any stage of frequency translation, the frequency band occupied. *Note:* For example, the output of a group multiplexer consists of a band of frequencies from 60 kHz to 108 kHz. This is the group-level baseband that results from combining 12 voice-frequency input channels, having a bandwidth of 4 kHz each, including guard bands. In turn, 5 groups are multiplexed into a super group having a baseband of 312 kHz to 552 kHz. This baseband, however, does not represent a group-level baseband. Ten super groups are in turn multiplexed into one master group, the output of which is a baseband that may be used to modulate a microwave-frequency carrier.

multiplexer (MUX). A device that combines multiple inputs into an aggregate signal to be transported via a single transmission channel. *Synonym* multiplexing equipment.

multiplexer/demultiplexer (muldem). A device that combines the functions of multiplexing and demultiplexing of digital signals. *Note:* The term *muldem* should not be confused with *modem*.

multiplex hierarchy. In frequency-division multiplexing, the rank of frequency bands occupied:

12 channels	group
5 groups (60 channels)	super group
5 super groups (300 channels)	master group (CCITT)
10 super groups (600 channels)	master group (U.S. standard)
6 U.S. master groups (3600 channels)	jumbo group

multiplexing (MUXing). The combining of two or more information channels onto a common transmission medium. *Note:* In electrical communications, the two basic forms of multiplexing are time-division multiplexing (TDM) and frequency-division multiplexing (FDM). In optical communications, the analog of FDM is referred to as wavelength-division multiplexing (WDM).

multiplexing equipment. *Synonym* multiplexer.

multiplex link encryption. Encryption in which a single cryptographic device is used to encrypt all of the data in a multiplexed link.

multipoint. *Synonym* multiple.

multipoint access. Access in which more than one terminal is supported by a single network termination.

multipoint circuit. A circuit that interconnects three or more separate points.

multipoint distribution service. A one-way domestic public radio service rendered on microwave frequencies from a fixed station transmitting (usually in an omnidirectional pattern) to multiple receiving facilities located at fixed points.

multipoint grounding system. Equipment bonded together and also bonded to the facility grounding system at the point nearest the equipment.

multipoint link. A data communications link that interconnects three or more terminals.

multiport repeater. In digital networking, an active device, having multiple input/output (I/O) ports, in which a signal introduced at the input of any port appears at the output of every port. *Note 1:* A multiport repeater usually performs regenerative functions, *i.e.*, it reshapes the digital signals. *Note 2:* Depending on the application, a multiport repeater may be designed not to repeat a signal back to the port from which it originated.

multiprocessing. 1. Simultaneous processing by two or more processors acting in concert. 2. The simultaneous execution of two or more computer programs or sequences of instructions by a single processor.

multiprocessor. A computer that has two or more processors that have common access to a main storage.

multiprogramming. A mode of operation that provides for the interleaved execution of two or more computer programs by a single processor.

multi-satellite link. A radio link between a transmitting Earth station and a receiving Earth station through two or more satellites, without any intermediate Earth station. A multi-satellite link comprises one uplink, one or more satellite-to-satellite links, and one downlink.

multitasking. The concurrent performance or interleaved execution of two or more tasks. *Synonym* concurrent operation.

mutually synchronized network. A network that has a synchronizing arrangement in which each clock in the network exerts a degree of control on all others.

mutual synchronization. Synchronization in which the frequency of the clock at a particular node is controlled by a weighted average of the timing on all signals received from neighboring nodes.

MUX. *Abbreviation for* multiplex, multiplexer. *See* multiplexing.

mw. *Abbreviation for* microwave.

NA. *Abbreviation for* numerical aperture.

nailed-up circuit. *Deprecated term. See* **dedicated circuit, permanent virtual circuit.**

NAK. *Acronym for* negative-acknowledge character.

NAK attack. In communications security systems, a security penetration technique that makes use of the negative-acknowledge transmission-control character and capitalizes on a potential weakness in a system that handles asynchronous transmission interruption in such a manner that the system is in an unprotected state against unauthorized access during certain periods.

narrative traffic. Traffic consisting of plain or encrypted messages written in a natural language and transmitted in accordance with standard formats and procedures. *Note:* Examples of narrative traffic include (a) messages that are placed on paper tape and transmitted via a teletypewriter (TTY), and on reception, are converted back to a printed page on another teletypewriter or teleprinter and (b) messages printed on a sheet of paper, transmitted via optical character recognition (OCR) equipment, and on reception, converted back to a printed page on a printer.

narrowband modem. A modem whose modulated output signal has an essential frequency spectrum that is limited to that which can be wholly contained within, and faithfully transmitted through, a voice channel with a nominal 4-kHz bandwidth. *Note:* High frequency (HF) modems are limited to operation over a voice channel with a nominal 3-kHz bandwidth.

narrowband radio voice frequency (NBRVF). In narrowband radio, the nominal 3-kHz bandwidth allocated for single channel radio that provides a transmission path for analog and quasi-analog signals.

narrowband signal. Any analog signal or analog representation of a digital signal whose essential spectral content is limited to that which can be contained within a voice channel of nominal 4-kHz bandwidth. *Note:* Narrowband radio uses a voice channel with a nominal 3-kHz bandwidth.

***n*-ary code.** A code that has *n* significant conditions, where n is a positive integer greater than 1. *Note 1:* The integer substituted for *n* indicates the specific number of significant conditions, *i.e.*, quantization states, in the code. For example, an 8-ary code has eight significant conditions and can convey three bits per code symbol. *Note 2:* A prefix that indicates an integer, *e.g.*, "bi," "tern," or "quater," may be used in lieu of a numeral, to produce "binary," "ternary," or "quaternary" (2, 3, and 4 states respectively).

***n*-ary signaling.** *See n*-ary code.

NATA. *Abbreviation for* North American Telecommunications Association.

National Command Authorities (NCA). The President and the Secretary of Defense or their duly deputized alternates or successors.

National Communications System (NCS). 1. The organization established by Section 1(a) of Executive Order No. 12472 to assist the President, the National Security Council, the Director of the Office of Science and Technology Policy, and the Director of the Office of Management and Budget, in the discharge of their national security emergency preparedness telecommunications functions. The NCS consists of both the telecommunications assets of the entities represented on the NCS Committee of Principals and an administrative structure consisting of the Executive Agent, the NCS Committee of Principals, and the Manager. 2. The telecommunications system that results from the technical and operational integration of the separate telecommunications systems of the several executive branch departments and agencies having a significant telecommunications capability.

National Coordinating Center (NCC) for Telecommunications. The joint telecommunications industry/Federal Government operation established by the National Communications System to assist in the initiation, coordination, restoration, and reconstitution of National Security or Emergency Preparedness (NS/EP) telecommunications services or facilities.

National Electric Code® (NEC). A standard that governs the use of electrical wire, cable, and fixtures, and electrical and optical communications cable installed in buildings. *Note:* The NEC was developed by the NEC Committee of the American National Standards Institute (ANSI), was sponsored by the National Fire Protection Association (NFPA), and is identified by the description ANSI/NFPA 70-XXXX, the last four digits representing the year of the NEC revision.

National Information Infrastructure (NII). A proposed, advanced, seamless web of public and private communications networks, interactive services, interoperable hardware and software, computers, databases, and consumer electronics to put vast amounts of information at users' fingertips. *Note:* NII includes more than just the physical facilities (more than the cameras, scanners, keyboards, telephones, fax machines, computers, switches, compact disks, video and audio tape, cable, wire, satellites, optical fiber transmission lines, microwave nets, switches, televisions, monitors, and printers) used to transmit, store, process, and display voice, data, and images; it encompasses a wide range of interactive functions, user-tailored services, and multimedia databases that are interconnected in a technology-neutral manner that will favor no one industry over any other. *Synonym* information superhighway.

National Security or Emergency Preparedness telecommunications. *See* NS/EP telecommunications.

National Television Standards Committee standard. *See* NTSC standard.

natural frequency. Of an antenna, the lowest frequency at which the antenna resonates without the addition of any inductance or capacitance.

nautical mile (nmi). A unit of distance used in navigation and based on the length of one minute of arc taken along a great circle. *Note 1:* Because the Earth is not a perfect sphere, various values have been assigned to the nautical mile. The value 1852 meters (6076.1 ft.) has been adopted internationally. *Note 2:* The nautical mile is frequently confused with the geographical mile, which is equal to 1 min of arc on the Earth's equator (6087.15 ft.).

NBH. *Abbreviation for* network busy hour. *See* busy hour.

NBRVF. *Abbreviation for* narrowband radio voice frequency.

NCA. *Abbreviation for* National Command Authorities.

NCC. *Abbreviation for* National Coordinating Center for Telecommunications.

NCS. *Abbreviation for* National Communications System, net control station.

near-end crosstalk. Crosstalk that is propagated in a disturbed channel in the direction opposite to the direction of propagation of a signal in the disturbing channel. *Note:* The terminals of the disturbed channel, at which the near-end crosstalk is present, and the energized terminal of the disturbing channel, are usually near each other.

near field. *Synonym* near-field region (def. #1).

near-field diffraction pattern. The diffraction pattern of an electromagnetic wave, which pattern is observed close to a source or aperture, as distinguished from a far-field diffraction pattern. *Note:* The pattern in the output plane is called the near-field radiation pattern. *Synonym* **Fresnel diffraction pattern.** *Contrast with* far-field diffraction pattern.

near-field region. 1. The close-in region of an antenna wherein the angular field distribution is dependent upon distance from the antenna. *Synonyms* near field, near zone. 2. In optical fiber communications, the region close to a source or aperture. *Note:* The diffraction pattern in this region typically differs significantly from that observed at infinity and varies with distance from the source.

near-field scanning. A technique for measuring the refractive-index profile of an optical fiber by using an extended source to illuminate an endface and measuring the point-by-point radiance at the exit face.

near real time. 1. Pertaining to the delay introduced, by automated data processing, between the occurrence of an event and the use of the processed data, *e.g.*, for display or feedback and control purposes. *Note 1:* For example, a near-real-time display depicts an event or situation as it existed at the current time less the processing time. *Note 2:* The distinction between near real time and real time is somewhat nebulous and must be defined for the situation at hand. *Contrast with* real time. 2. Pertaining to the timeliness of data or information which has been delayed by the time required for electronic communication and automatic data processing. This implies that there are no significant delays.

near-vertical-incidence skywave. In radio propagation, a wave that is reflected from the ionosphere at a nearly vertical angle and that is used in short-range communications to reduce the area of the skip zone and thereby improve reception beyond the limits of the ground wave.

near zone. *Synonym* near-field region (def. #1).

NEC. *Abbreviation for* National Electric Code®.

necessary bandwidth. For a given class of emission, the width of the frequency band which is just sufficient to ensure the transmission of information at the rate and with the quality required under specified conditions. *Note:* Emissions useful for the adequate functioning of the receiving equipment, *e.g.*, the emission corresponding to the carrier of reduced carrier systems, must be included in the necessary bandwidth.

negation circuit. *Deprecated synonym for* inverter.

negative-acknowledge character (NAK). A transmission control character sent by a station as a negative response to the station with which the connection has been set up. *Note 1:* In binary synchronous communication protocol, the NAK is used to indicate that an error was detected in the previously received block and that the receiver is ready to accept retransmission of that block. *Note 2:* In multipoint systems, the NAK is used as the not-ready reply to a poll.

negative feedback. *See* feedback (def. #1).

negative justification. *Synonym* de-stuffing.

negative pulse stuffing. *Synonym* de-stuffing.

***n*-entity.** An active element in the *n*-th layer of the Open Systems Interconnection—Reference Model (OSI-RM) that (a) interacts directly with elements, *i.e.*, entities, of the layer immediately above or below the *n*-th layer, (b) is defined by a unique set of rules, *i.e.*, syntax, and information formats, including data and control formats, and (c) performs a defined set of functions. *Note 1:* The *n* refers to any one of the 7 layers of the OSI-RM. *Note 2:* In an existing layered open system, the *n* may refer to any given layer in the system. *Note 3:* Layers are conventionally numbered from the lowest, *i.e.,* the physical layer, to the highest, so that the $n + 1$ layer is above the *n*-th layer and the $n - 1$ layer is below.

NEP. *Abbreviation for* noise equivalent power.

neper (Np). A unit used to express ratios, such as gain, loss, and relative values. *Note 1:* The neper is analogous to the decibel, except that the Naperian base 2.718281828. . . is used in computing the ratio in nepers. *Note 2:* The value in nepers, *Np*, is given by $Np = \ln(x_1/x_2)$, where x_1 and x_2 are the values of interest, and ln is the natural logarithm, *i.e.,* logarithm to the base e. *Note 3:* One neper (Np) = 8.686 dB, where 8.686 = 20/(ln 10). *Note 4:* The neper is often used to express voltage and current ratios, whereas the decibel is usually used to express power ratios. *Note 5:* Like the dB, the Np is a dimensionless unit. *Note 6:* The ITU recognizes both units.

nested command menu. A command menu within another command menu. *See* command menu.

net. *Synonym* communications net.

net control station (NCS). 1. A radio station that performs net control functions, such as controlling traffic and enforcing operational discipline. 2. [A] terminal in a secure telecommunications net responsible for distributing key in electronic form to the members of the net.

net gain. The overall gain of a transmission circuit. *Note 1:* Net gain is measured by applying a test signal at an appropriate power level *(see Note 5)* at the input port of a circuit and measuring the power delivered at the output port. The net gain in dB is calculated by taking 10 times the logarithm of the ratio of the output power to the input power. *Note 2:* The net gain expressed in dB may be positive or negative. *Note 3:* If the net gain expressed in dB is negative, it is also called the "net loss." *Note 4:* If the net gain is expressed as a ratio, and the ratio is less than unity, a net loss is indicated. *Note 5:* The test signal must be chosen so that its power level is within the usual operating range of the circuit being tested.

net loss. The overall loss of a transmission circuit.

net loss variation. The maximum change in net loss occurring in a specified portion of a communication system during a specified period.

net operation. The operation of an organization of stations capable of direct communication on a common channel or frequency. *Note:* Net operations (a) allow participants to conduct ordered conferences among participants who usually have common information needs or related functions to perform, (b) are characterized by adherence to standard formats and procedures, and (c) are responsive to a common supervisory station, called the "*net control station,*" which permits access to the net and maintains net operational discipline.

net radio interface (NRI). An interface between a single-channel radio station (usually in a radio net) and switched communications systems.

network. 1. An interconnection of three or more communicating entities. 2. An interconnection of usually passive electronic components that performs a specific function (which is usually limited in scope), *e.g.,* to simulate a transmission line or to perform a mathematical function such as integration or differentiation. *Note:* A network may be part of a larger circuit.

network administration. A group of network management functions that (a) provide support services, (b) ensure that the network is used efficiently, and (c) ensure prescribed service-quality objectives are met. *Note:* Network administration may include activities such as network address assignment, assignment of routing protocols and routing table configuration, and directory service configuration.

network architecture. 1. The design principles, physical configuration, functional organization, operational procedures, and data formats used as the bases for the design, construction, modification, and operation of a communications network. 2. The structure of an existing communications network, including the physical configuration, facilities, operational structure, operational procedures, and the data formats in use.

network busy hour (NBH). *See* busy hour.

network connectivity. The topological description of a network that specifies, in terms of circuit termination locations and quantities, the interconnection of the transmission nodes.

network control program (NCP). In a switch or network node, software designed to store and forward frames between nodes. *Note:* An NCP may be used in local area networks or larger networks.

network element (NE). In integrated services digital networks, a piece of telecommunications equipment that provides support or services to the user.

network engineering. 1. In telephony, the discipline concerned with (a) determining internetworking service requirements for switched networks, and (b) developing and implementing hardware and software to meet them. 2. In computer science, the discipline of hardware and software engineering to accomplish the design goals of a computer network. 3. In radio communications, the discipline concerned with developing network topologies.

Network File System. *See* NFS.

network interface. 1. The point of interconnection between a user terminal and a private or public network. 2. The point of interconnection between the public switched network and a privately owned terminal. *Note: Code of Federal Regulations,* Title 47, part 68, stipulates the interface parameters. 3. The point of interconnection between one network and another network.

network interface card (NIC). A network interface device (NID) in the form of circuit card that is installed in an expansion slot of a computer, to provide network access. *Note:* Examples of NICs are cards that interface a computer with an Ethernet LAN and cards that interface a computer with an FDDI ring network.

network interface device (NID). 1. A device that performs interface functions, such as code conversion, protocol conversion, and buffering, required for communications to and from a network. 2. A device used primarily within a local area network (LAN) to allow a number of independent devices, with varying protocols, to communicate with each other. *Note 1:* An NID converts each device protocol into a common transmission protocol. *Note 2:* The transmission protocol may be chosen to accommodate directly a number of the devices used within the network without the need for protocol conversion for those devices by the NID. *Synonym* **network interface unit.**

network interface unit (NIU). *Synonym* network interface device.

network inward dialing (NID). *Synonym* direct inward dialing.

Network Layer. *See* Open Systems Interconnection–Reference Model.

network management. The execution of the set of functions required for controlling, planning, allocating, deploying, coordinating, and monitoring the resources of a telecommunications network, including performing functions such as initial network planning, frequency allocation, predetermined traffic routing to support load balancing, cryptographic key distribution authorization, configuration management, fault management, security management, performance management, and accounting management. *Note:* Network management does not include user terminal equipment.

network manager. In network management, the entity that initiates requests for management information from managed systems or receives spontaneous management-related notifications from managed systems.

network operating system (NOS). Software that (a) controls a network and its message (*e.g.*, packet) traffic, and queues, (b) controls access by multiple users to network resources such as files, and (c) provides for certain administrative functions, including security. *Note 1:* A network operating system is most frequently used with local area networks and wide area networks, but could also have application to larger network systems. *Note 2:* The upper 5 layers of the OSI—Reference Model provide the foundation upon which many network operating systems are based.

network outward dialing (NOD). *Synonym* direct outward dialing.

network terminal number (NTN). In the CCITT International X.121 format, the sets of digits that comprise the complete address of the data terminal end point. *Note:* For an NTN that is not part of a national integrated numbering format, the NTN is the 10 digits of the CCITT X.25 14-digit address that follow the Data Network Identification Code (DNIC). When part of a national integrated numbering format, the NTN is the 11 digits of the CCITT X.25 14-digit address that follow the DNIC.

network terminating interface (NTI). *Synonym for* demarcation point.

network termination. Network equipment that provides functions necessary for network operation of ISDN access protocols. *Note:* Network termination provides functions essential for transmission services.

network termination 1 (NT1). In Integrated Services Digital Networks (ISDN), a functional grouping of customer-premises equipment that includes functions that may be regarded as belonging to OSI Layer 1, *i.e.,* functions associated with ISDN electrical and physical terminations on the user premises. *Note:* The

NT1 forms a boundary to the network and may be controlled by the provider of the ISDN services.

network termination 2 (NT2). In Integrated Services Digital Networks (ISDN), an intelligent device that may include functionality for OSI Layers 1 through 3 (dependent on individual systems requirements).

network topology. The specific physical, *i.e.,* real, or logical, *i.e.,* virtual, arrangement of the elements of a network. *Note 1:* Two networks have the same topology if the connection configuration is the same, although the networks may differ in physical interconnections, distances between nodes, transmission rates, and/or signal types. *Note 2:* The common types of network topology are illustrated *[refer to the figure on this page]* and defined in alphabetical order below:

bus topology. A network topology in which all nodes, *i.e.,* stations, are connected together by a single bus.

fully connected topology. A network topology in which there is a direct path (branch) between any two nodes. *Note:* In a fully connected network with n nodes, there are n(n−1)/2 direct paths, *i.e.,* branches. *Synonym* fully connected mesh network.

hybrid topology. A combination of any two or more network topologies. *Note 1:* Instances can occur where two basic network topologies, when connected together, can still retain the basic network character, and therefore not be a hybrid network. For example, a tree network connected to a tree network is still a tree network. Therefore, a hybrid network accrues only when two basic networks are connected and the resulting network topology fails to meet one of the basic topology definitions. For example, two star networks connected together exhibit hybrid network topologies. *Note 2:* A hybrid topology always accrues when two different basic network topologies are connected.

linear topology. *See* bus topology.

mesh topology. A network topology in which there are at least two nodes with two or more paths between them.

ring topology. A network topology in which every node has exactly two branches connected to it.

star topology. A network topology in which peripheral nodes are connected to a central node, which rebroadcasts all transmissions received from any peripheral node to all peripheral nodes on the network, including the originating node. *Note 1:* All peripheral nodes may thus communicate with all others by transmitting to, and receiving from, the central node only. *Note 2:* The failure of a transmission line, *i.e.,* channel, linking any peripheral node to the central node will result in the isolation of that peripheral node from all others. *Note 3:* If the star central node is passive, the originating node must be able to tolerate the reception of an echo of its own transmission, delayed by the two-way transmission time, *i.e.,* to and from the central node, plus any delay generated in the central node. An active star network has an active central node that usually has the means to prevent echo-related problems.

tree topology. A network topology that, from a purely topologic viewpoint, resembles an interconnection of star networks in that individual peripheral nodes are required to transmit to and receive from one other node only, toward a central node, and are not required to act as repeaters or regenerators. *Note 1:* The function of the central node may be distributed. *Note 2:* As in the conventional star network, individual nodes may thus still be isolated from the network by a single-point failure of a transmission path to the node. *Note 3:* A single-point failure of a transmission path within a distributed node will result in partitioning two or more stations from the rest of the network.

network utility. An internetwork administrative signaling mechanism in the call control procedure between packet switching public data networks.

neutral. 1. In ac power distribution, the conductor that (a) is intentionally grounded on the supply side of the service disconnect and (b) provides a current return path for ac power currents. 2. In three-phase ac "Y," *i.e.,* wye, power distribution, the low-potential fourth wire that conducts only that current required to achieve electrical balance, *i.e.,* to provide a return path for any current imbalance among the three phases.

neutral direct-current telegraph system. A telegraph system in which (a) current flows during marking intervals and no current flows during spacing intervals for the transmission of signals over a line, and (b) the direction of current flow is immaterial. *Synonyms* single-current system, single-current transmission system, single-Morse system.

neutral ground. An intentional ground applied to the neutral conductor or neutral point of a circuit, transformer, machine, apparatus, or system.

neutral operation. A method of teletypewriter operation in which marking signals are formed by current pulses of one polarity, either positive or negative, and spacing signals are formed by reducing the current to zero or nearly zero.

neutral relay. A relay in which the direction of movement of the armature does not depend upon the direction of the current in the circuit controlling the armature.

new customer premises equipment. All customer premises equipment not in service or in the inventory of a regulated telephone utility as of December 31, 1982.

NF. *Abbreviation for* noise figure.

NFS. *Abbreviation for* Network File System. A proprietary distributed file system that is widely used by TCP/IP vendors. *Note:* NFS allows different computer systems to share files, and uses user datagram protocol (UDP) for data transfer.

***n*-function.** A defined action performed by an *n*-entity. *Note:* An *n*-function may be (a) a single action, *i.e.*, a primitive function, or (b) a set of actions.

nibble. Part of a byte, usually half of a byte. *(Obsolete)*

NIC. *Abbreviation for* network interface card.

NID. *Acronym for* network interface device, network inward dialing.

NII. *Abbreviation for* National Information Infrastructure.

nine-hundred (900) service. A telephone service via which the caller may access information on a charge-per-call or charge-per-time basis.

NIU. *Abbreviation for* network interface unit. *See* network interface device.

nmi. *Abbreviation for* nautical mile.

NOD. *Acronym for* network outward dialing.

nodal clock. The principal clock or alternate clock located at a particular node that provides the timing reference for all major functions at that node.

nodal point. *Synonym* node (def. #1).

node. 1. In network topology, a terminal of any branch of a network or an interconnection common to two or more branches of a network. *Synonyms* junction point, nodal point. 2. In a switched network, one of the switches forming the network backbone. *Note:* A node may also include patching and control facilities. 3. A technical control facility (TCF). 4. A point in a standing or stationary wave at which the amplitude is a minimum. *In this sense, synonym* null.

noise. 1. An undesired disturbance within the frequency band of interest; the summation of unwanted or disturbing energy introduced into a communications system from man-made and natural sources. 2. A disturbance that affects a signal and that may distort the information carried by the signal. 3. Random variations of one or more characteristics of any entity such as voltage, current, or data. 4. A random signal of known statistical properties of amplitude, distribution, and spectral density. 5. *Loosely,* any disturbance tending to interfere with the normal operation of a device or system.

noise current. 1. Interfering and unwanted electrical currents in a device or system. 2. In optical communications, the rms component of the optical detector output electrical current with no incoming signal present.

noise equivalent power (NEP). At a given data-signaling rate or modulation frequency, operating wavelength, and effective noise bandwidth, the radiant power that produces a signal-to-noise ratio of unity at the output of a given optical detector. *Note 1:* Some manufacturers and authors define NEP as the minimum detectable power per square root bandwidth. When defined this way, NEP has the units of watts per $(\text{hertz})^{1/2}$. Therefore, the term is a misnomer, because the units of power are watts. *Note 2:* Some manufacturers define NEP as the radiant power that produces a signal-to-dark-current noise ratio of unity. The NEP measurement is valid only if the dark-current noise dominates the noise level.

noise factor. *Synonym* noise figure.

noise figure (NF). The ratio of the output noise power of a device to the portion thereof attributable to thermal noise in the input termination at standard noise temperature (usually 290 K). *Note:* The noise figure is thus the ratio of actual output noise to that which would remain if the device itself did not introduce noise. In heterodyne systems, output noise power includes spurious contributions from image-frequency transformation, but the portion attributable to thermal noise in the input termination at standard noise temperature includes only that which appears in the output via the principal frequency transformation of the system, and excludes that which appears via the image frequency transformation. *Synonym* noise factor.

noise level. The noise power, usually relative to a reference. *Note:* Noise level is usually measured in dB for relative power or picowatts for absolute power. A suffix is added to denote a particular reference base or specific qualities of the measurement. Examples of noise-level measurement units are dBa, dBa(F1A), dBa(HA1), dBa0, dBm, dBm(psoph), dBm0, dBm0P, dBrn, dBrnC, dBrn(f_1-f_2), dBrn(144-line), pW, pWp, and pWp0.

noise power. 1. The power generated by a random electromagnetic process. 2. Interfering and unwanted power in an electrical device or system. 3. In the acceptance testing of radio transmitters, the mean power supplied to the antenna transmission line by a radio transmitter when loaded with noise having a Gaussian amplitude-vs.-frequency distribution.

noise power density. The noise power in a bandwidth of 1 Hz, *i.e.*, the noise power per hertz at a point in a noise spectrum. *Note:* The noise-power density of the internal noise that is contributed by a receiving system to an incoming signal is expressed as the product of Boltzmann's constant, k, and the equivalent noise temperature, T_n. Thus, the noise-power density is often expressed simply as k*T*. *Synonym* k*T*.

noise suppression. 1. Reduction of the noise power level in electrical circuits. 2. The process of automatically reducing the noise output of a receiver during periods when no carrier is being received. *Contrast with* squelch.

noise temperature. At a pair of terminals, the temperature of a passive system having an available noise power per unit bandwidth at a specified frequency equal to that of the actual terminals of a network. *Note:* The noise temperature of a simple resistor is the actual temperature of that resistor. The noise temperature of a diode may be many times the actual temperature of the diode.

noise voltage. 1. Interfering and unwanted voltage in an electronic device or system. 2. In optical communications, the rms component of the optical detector output electrical voltage with no incoming signal present.

noise weighting. A specific amplitude-vs.-frequency characteristic that permits a measuring set to give numerical readings that approximate the interfering effects to any listener using a particular class of telephone instrument. *Note 1:* Noise weighting measurements are made in lines terminated either by the measuring set or the class of instrument. *Note 2:* The most widely used noise weightings were established by agencies concerned with public telephone service, and are based on characteristics of specific commercial telephone instruments, representing successive stages of technological development. The coding of commercial apparatus appears in the nomenclature of certain weightings. The same weighting nomenclature and units are used in military versions of commercial noise measuring sets.

noise window. A notch, *i.e.,* a dip, in the noise frequency spectrum characteristic of a device, such as a transmitter, receiver, channel, or amplifier, from external sources or internal sources. *Note:* The noise window is usually represented as a band of lower amplitude noise in a wider band of higher amplitude noise.

noisy black. 1. In facsimile or display systems, such as television, a nonuniformity in the black area of the image, *i.e.,* document or picture, caused by the presence of noise in the received signal. 2. A signal or signal level that is supposed to represent a black area on the object, but has a noise content sufficient to cause the creation of noticeable white spots on the display surface or record medium.

noisy white. 1. In facsimile or display systems, such as television, a nonuniformity in the white area of the image, *i.e.,,* document or picture, caused by the presence of noise in the received signal. 2. A signal or signal level that is supposed to represent a white area on the object, but has a noise content sufficient to cause the creation of noticeable black spots on the display surface or record medium.

nominal bandwidth. The widest band of frequencies, inclusive of guard bands, assigned to a channel. *Note:* Nominal bandwidth should not be confused with the terms *"necessary bandwidth," "occupied bandwidth,"* or *"rf bandwidth."*

nominal bit stuffing rate. The rate at which stuffing bits are inserted when both the input and output bit rates are at their nominal values.

nominal linewidth. In facsimile systems, the average separation between centers of adjacent scanning or recording lines.

nonassociated common-channel signaling. A form of common-channel signaling where the signaling channel serves one or more trunk groups, at least one of which terminates at a point other than the signal transfer point at which the signaling channel terminates.

nonblocking switch. A switch that has enough paths across it that an originated call can always reach an available line without encountering a busy condition.

non-call associated signaling (NCAS). Signaling that is independent of an end-to-end bearer connection, including support for the functions of registration, authentication, and validation.

noncentralized operation. Operation that uses a control discipline for multipoint data communication links in which transmission may be between tributary stations or between the control station and tributary stations.

noncircularity. *Synonym* **ovality.**

noncritical technical load. Of the total technical load at a facility during normal operation, the part that is not required for synchronous operation.

nonerasable storage. *Synonym* read-only memory.

non-fixed access. In personal communications service (PCS), terminal access to a network in which there is no set relationship between a terminal and the access interface. *Note:* The access interface and the terminal each has its own separate "identifiers." The terminal may be moved from one access interface to another while maintaining the terminal's unique identity.

nonlinear distortion. Distortion caused by a deviation from a linear relationship between specified input and output parameters of a system or component.

nonlinear scattering. Direct conversion of a photon from one wavelength to one or more other wavelengths. *Note 1:* In an optical fiber, nonlinear scattering is usually not important below the threshold irradiance for stimulated nonlinear scattering. *Note 2:* Examples of nonlinear scattering are Raman and Brillouin scattering.

nonloaded twisted pair. A twisted pair that has no intentionally added inductance.

nonoperational load. Administrative, support, and housing power requirements. *Synonym* **utility load.**

nonresonant antenna. *Synonym* aperiodic antenna.

non-return-to-zero (NRZ). A code in which "1s" are represented by one significant condition and "0s" are represented by another, with no neutral or rest condition, such as a zero amplitude in amplitude modulation (AM), zero phase shift in phase-shift keying (PSK), or mid-frequency in frequency-shift keying (FSK). *Note 1: Contrast with* Manchester code, return-to-zero. *Note 2:* For a given data signaling rate, *i.e.,* bit rate, the NRZ code requires only one-half the bandwidth required by the Manchester code.

non-return-to-zero change-on-ones (NRZ1). A code in which "1s" are represented by a change in a significant condition and "0s" are represented by no change.

non-return-to-zero mark (NRZ-M). A binary encoding scheme in which a signal parameter, such as electric current or voltage, undergoes a change in a significant condition or level every time that a "one" occurs, but when a "zero" occurs, it remains the same, *i.e.*, no transition occurs. *Note 1:* The transitions could also occur only when "zeros" occur and not when "ones" occur. If the significant condition transition occurs on each "zero," the encoding scheme is called "non-return-to-zero space" (NRZ-S). *Note 2:* NRZ-M and NRZ-S signals are technically interchangeable; *i.e.,* one is the logical "NOT" (inverse) of the other. It is necessary for the receiver to have prior knowledge of which scheme is being used. Without such knowledge, it is impossible for the receiver to interpret the data stream correctly; *i.e.,* its output may be the correct data stream or the logical inverse of the correct data stream. *Contrast with* non-return-to-zero space. *Synonyms* conditioned baseband representation, differentially encoded baseband, non-return-to-zero one (NRZ-1), NRZ-B.

non-return-to-zero one (NRZ-1). *Synonym* non-return-to-zero mark.

non-return-to-zero space (NRZ-S). A binary encoding scheme in which a signal parameter, such as electric current or voltage, undergoes a change in a significant condition or level every time that a "zero" occurs, but when a "one" occurs, it remains the same, *i.e.*, no transition occurs. *Note 1:* The transitions could also occur only when "ones" occur and not when "zeros" occur. If the significant condition transition occurs on each "one," the encoding scheme is called "non-return to zero mark" (NRZ-M). *Note 2:* NRZ-S and NRZ-M signals are technically interchangeable; *i.e.,* one is the logical "NOT" (inverse) of the other. It is necessary for the receiver to have prior knowledge of which scheme is being used. Without such knowledge, it is impossible for the receiver to interpret the data stream correctly; *i.e.,* its output may be the correct data stream or the logical inverse of the correct data stream. *Contrast* *with* non-return-to-zero mark. *Synonym* non-return-to-zero.

nonshifted fiber. *Synonym* dispersion-unshifted fiber.

nonsynchronous data transmission channel. A data transmission channel in which separate timing information is not transferred between the data terminal equipment (DTE) and the data circuit terminating equipment (DCE).

nonsynchronous network. *Synonym* asynchronous network.

nonsynchronous system. *See* asynchronous transmission.

nonsynchronous transmission. *See* asynchronous transmission.

nontechnical load. Of the total operational load at a facility during normal operation, the part used for support purposes, such as general lighting, heating, air-conditioning, and ventilating equipment.

nontransparent mode. A mode of operating a data transmission system in which control characters are treated and interpreted as such, rather than simply as data or text bits in a bit.

normalized frequency (*V*). 1. In an optical fiber, a dimensionless quantity, *V*, given by

$$V = \frac{2\pi a}{\lambda}\sqrt{n_1^2 - n_2^2}\ , V = \frac{2\pi a}{\lambda}\sqrt{n_1^2 - n_2^2}\ ,$$

where a is the core radius, λ is the wavelength in vacuum, n_1 is the maximum refractive index of the core, and n_2 is the refractive index of the homogeneous cladding. *Note 1:* In multimode operation of an optical fiber having a power-law refractive index profile, the approximate number of bound modes, *i.e.*, the mode volume, is given by

$$\frac{V^2}{2}\left(\frac{g}{g+2}\right),$$

where *V* is the normalized frequency greater than 5 and *g* is the profile parameter. *Note 2:* For a step index fiber, the mode volume is given by $V^2/2$. For single-mode operation, $V < 2.405$. *Synonym* *V* number. 2. The ratio between an actual frequency and a reference value. 3. The ratio between an actual frequency and its nominal value.

notch. In a relatively wide band of frequencies, not necessarily of uniform amplitude, a narrow band of frequencies having relatively low amplitudes.

notched filter. *Synonym* band-stop filter.

notched noise. Noise from which a narrow band of frequencies has been removed. *Note:* Notched noise is usually used for testing devices or circuits.

not-ready condition. At the data terminal equipment/data circuit-terminating equipment (DTE/DCE) interface, a steady-state condition that indicates that the DCE is not ready to accept a call-request signal or that the DTE is not ready to accept an incoming call.

Np. *Abbreviation for* neper.

NRI. *Abbreviation for* net radio interface.

NRZ. *Abbreviation for* non-return-to-zero.

NRZ-M. *Synonym* non-return-to-zero mark.

NS/EP telecommunications. *Abbreviation for* National Security or Emergency Preparedness telecommunications. Telecommunications services that are used to maintain a state of readiness or to respond to and manage any event or crisis (local, national, or international) that causes or could cause injury or harm to the population, damage to or loss of property, or degrade or threaten the national security or emergency preparedness posture of the United States.

***n*-sequence.** A pseudorandom binary sequence of *n* bits that (a) is the output of a linear shift register and (b) has the property that, if the shift register is set to any nonzero state and then cycled, a pseudorandom binary sequence of a maximum of $n = 2^m - 1$ bits will be generated, where m is the number of stages, *i.e.*, the number of bit positions in the register, before the shift register returns to its original state and the *n*-bit output sequence repeats. *Note:* The register may be used to control the sequence of frequencies for a frequency-hopping spread spectrum transmission system.

NTI. *Abbreviation for* network terminating interface.

NTN. *Abbreviation for* network terminal number.

NTSC standard. *Abbreviation for* National Television Standards Committee standard. The North American standard (525-line interlaced raster-scanned video) for the generation, transmission, and reception of television signals. *Note 1:* In the NTSC standard, picture information is transmitted in vestigial-sideband AM and sound information is transmitted in FM. *Note 2:* In addition to North America, the NTSC standard is used in Central America, a number of South American countries, and some Asian countries, including Japan. *Contrast with* PAL, PAL-M, SECAM.

nuclear hardness. 1. An expression of the extent to which the performance of a system, facility, or device is expected to degrade in a given nuclear environment. 2. The physical attributes of a system or component that will allow survival in an environment that includes nuclear radiation and electromagnetic impulses (EMI). *Note 1:* Nuclear hardness may be expressed in terms of either susceptibility or vulnerability. *Note 2:* The extent of expected performance degradation (*e.g.,* outage time, data lost, and equipment damage) must be defined or specified. The environment (*e.g.,* radiation levels, overpressure, peak velocities, energy absorbed, and electrical stress) must be defined or specified. 3. The physical attributes of a system or component that will allow a defined degree of survivability in a given environment created by a nuclear weapon. *Note:* Nuclear hardness is determined for specified or actual quantified environmental conditions and physical parameters, such as peak radiation levels, overpressure, velocities, energy absorbed, and electrical stress. It is achieved through design specifications and is verified by test and analysis techniques.

null. 1. In an antenna radiation pattern, a zone in which the effective radiated power is at a minimum relative to the maximum effective radiated power of the main beam. *Note 1:* A null often has a narrow directivity angle compared to that of the main beam. Thus, the null is useful for several purposes, such as radio navigation and suppression of interfering signals in a given direction. *Note 2:* Because there is reciprocity between the transmitting and receiving characteristics of an antenna, there will be corresponding nulls for both the transmitting and receiving functions. 2. A dummy letter, letter symbol, or code group inserted in an encrypted message to delay or prevent its solution, or to complete encrypted groups for transmission or transmission security purposes. 3. In database management systems, a special value assigned to a row or a column indicating either unknown values or inapplicable usage. 4. *Synonym* node (def. #4).

null character (NUL). In transmission systems, a control character (a) that is used to accomplish media-fill stuffing or a time-fill stuffing in storage device or in a data transmission line and (b) that may be inserted and removed from a series of characters without affecting the meaning of the series. *Note:* The null character may affect the control of equipment or the format of messages.

numerical aperture (*NA*). 1. The sine of the vertex angle of the largest cone of meridional rays that can enter or leave an optical system or element, multiplied by the refractive index of the medium in which the vertex of the cone is located. *Note:* The *NA* is generally measured with respect to an object or image point and will vary as that point is moved. 2. For an optical fiber in which the refractive index decreases monotonically from n_1 on the axis to n_2 in the cladding, an expression of the extent of the fiber's ability to accept, in its bound modes, non-normal incident rays, given by $NA = (n_1^2 - n_2^2)^{1/2}$. *Note:* In multimode fibers, the term *equilibrium numerical aperture* is sometimes used. This refers to the numerical aperture with respect to the extreme exit angle of a ray emerging from a fiber in which equilibrium mode distribution has been established. 3. *Colloquially,* the sine of the radiation or acceptance angle of an optical fiber, multiplied by the refractive index of the material in contact with the exit or entrance face. *Note:* This usage is approximate and imprecise, but is often encountered.

numerical aperture loss. A loss of optical power that occurs at a splice or a pair of mated connectors when the numerical aperture of the "transmitting" fiber exceeds that of the "receiving" fiber, even if the cores are precisely the same diameter and are perfectly aligned. *Note 1:* The higher numerical aperture of the transmitting fiber means that it emits a larger cone of light than the receiving fiber is capable of accepting, resulting in a coupling loss. *Note 2:* In the opposite case of numerical aperture mismatch, where the transmitting fiber has the lower numerical aperture, no numerical aperture loss occurs, because the receiving fiber is capable of accepting light from any bound mode of the transmitting fiber.

***n*-unit code.** A code in which the signals or groups of digits that represent coded items, such as characters, have the same number of signal elements or digits, namely *n* elements or digits, where *n* may be any positive integer. *Note:* An example of an *n*-unit code is the 7-unit code (8-unit with parity) ASCII code. Each character is represented by a pattern of 7 binary digits. The units may also be characters or other special signs.

***n*-user.** In the ISO Open Systems Interconnection—Reference Model (OSI—RM), an *n*+1 entity that uses the services of the *n*-layer, and below, to communicate with another *n*+1 entity. *Note:* If *n* identifies a specific or a reference level, the *n*+1 layer is the layer above the n layer and the *n*–1 layer is the layer below the *n* layer. Thus, the *n*+2 layer is two layers above the *n* layer.

NVIS. *Abbreviation for* near vertical-incidence skywave.

NXX code. In the North American direct distance dialing numbering plan, a central office code of three digits that designates a particular central office or a given 10,000-line unit of subscriber lines; "N" is any number from 2 to 9, and "X" is any number from 0 to 9.

Nyquist interval. The maximum time interval between equally spaced samples of a signal that will enable the signal waveform to be completely determined. *Note 1:* The Nyquist interval is equal to the reciprocal of twice the highest frequency component of the sampled signal. *Note 2:* In practice, when analog signals are sampled for the purpose of digital transmission or other processing, the sampling rate must be more frequent than that defined by Nyquist's theorem, because of quantization error introduced by the digitizing process. The required sampling rate is determined by the accuracy of the digitizing process.

Nyquist rate. The reciprocal of the Nyquist interval, *i.e.,* the minimum theoretical sampling rate that fully describes a given signal, *i.e.,* enables its faithful reconstruction from the samples. *Note:* The actual sampling rate required to reconstruct the original signal will be somewhat higher than the Nyquist rate, because of quantization errors introduced by the sampling process.

Nyquist's theorem. A theorem, developed by H. Nyquist, which states that an analog signal waveform may be uniquely reconstructed, without error, from samples taken at equal time intervals. The sampling rate must be equal to, or greater than, twice the highest frequency component in the analog signal. *Synonym* sampling theorem.

object. 1. In image processing, a sub-region of an image that is perceived as a single entity. *Note:* An image can contain more than one object. 2. In facsimile systems, the image, the likeness of which is to be transmitted.

object persistence. In a video display, distortion wherein the entirety of some object (or objects) that appeared in a previous frame (and that should no longer appear) remain in the current frame and in subsequent frames as a faded image or as an outline.

object retention. In a video display, distortion in which a fragment of an object that appeared in a previous frame (and should no longer appear) remains in the current and subsequent video frames.

OCC. *Abbreviation for* other common carrier.

occupancy. For equipment, such as a circuit or a switch, the ratio of the actual time in use to the available time during a 1-hour period. *Note 1:* Occupancy is usually expressed in percent. *Note 2:* Occupancy may be plotted versus time of day. *Synonym* usage.

occupied bandwidth. The width of a frequency band such that, below the lower and above the upper frequency limits, the mean powers emitted are each equal to a specified percentage *B*/2 of the total mean power of a given emission. Unless otherwise specified by the CCIR for the appropriate class of emission, the value of $B/2$ should be taken as 0.5%. *Note 1:* The percentage of the total power outside the occupied bandwidth is represented by B. *Note 2:* In some cases, *e.g.*, multichannel frequency-division multiplexing systems, use of the 0.5% limits may lead to certain difficulties in the practical application of the definition of occupied and necessary bandwidth; in such cases, a different percentage may prove useful.

oceanographic data interrogating station. A station in the maritime mobile service the emissions of which are used to initiate, modify or terminate functions of equipment directly associated with an oceanographic data station, including the station itself.

oceanographic data station. A station in the maritime mobile service located on a ship, buoy, or other sensor platform the emissions of which are used for transmission of oceanographic data.

OCR. *Abbreviation for* optical character reader, optical character recognition.

octet. A byte of eight binary digits usually operated upon as an entity.

octet alignment. The configuration of a field composed of an integral number of octets. *Note:* If the field is not divisible by eight, bits (usually zeros) are added to either the first octet (left justification) or the last octet (right justification).

OD. *Abbreviation for* optical density.

odd-even check. *Synonym* parity check.

odd parity. *See* parity, parity check.

OFC. *Abbreviation for* optical fiber, conductive. *Note:* OFC is the designation given by the National Fire

Protection Association (NFPA) to interior fiber-optic cables which contain at least one electrically conductive, non-current-carrying component, such as a metallic strength member or vapor barrier, and which are not certified for use in plenum or riser applications.

OFCP. *Abbreviation for* optical fiber, conductive, plenum. *Note:* OFCP is the designation given by the National Fire Protection Association (NFPA) to interior fiber-optic cables which contain at least one electrically conductive, non-current-carrying component such as a metallic strength member or vapor barrier, and which are certified for use in plenum applications.

OFCR. *Abbreviation for* optical fiber, conductive, riser. *Note:* OFCR is the designation given by the National Fire Protection Association (NFPA) to interior fiber-optic cables which contain at least one electrically conductive, non-current-carrying component such as a metallic strength member or vapor barrier, and which are certified for use in riser applications.

off-axis optical system. An optical system in which the optical axis of the aperture is not coincident with the mechanical center of the aperture. *Note:* The principal applications of off-axis optical systems are to avoid obstruction of the primary aperture by secondary optical elements, instrument packages, or sensors, and to provide ready access to instrument packages or sensors at the focus. The engineering tradeoff of an off-axis optical system is an increase in image aberrations.

off-hook. 1. In telephony, the condition that exists when an operational telephone instrument or other user instrument is in use, *i.e.,* during dialing or communicating. *Note:* Off-hook originally referred to the condition that prevailed when the separate earpiece, *i.e.*, receiver, was removed from its switchhook, which extended from a vertical post that also supported the microphone, and which connected the instrument to the line when not depressed by the weight of the receiver. 2. One of two possible signaling states, such as tone or no tone and ground connection versus battery connection. *Note:* If off-hook pertains to one state, on-hook pertains to the other. 3. The active state, *i.e.*, closed loop, of a subscriber or PBX user loop. 4. An operating state of a communications link in which data transmission is enabled either for (a) voice or data communications or (b) network signaling.

off-hook service. *Synonym* hotline.

off-hook signal. In telephony, of a circuit, a signal indicating seizure, request for service, or a busy condition.

office classification. Prior to divestiture, numbers that were assigned to offices according to their hierarchical function in the U.S. public switched telephone network. *Note 1:* The following class numbers are used:

Class 1. Regional Center (RC)

Class 2. Sectional Center (SC)

Class 3. Primary Center (PC)

Class 4. Toll Center (TC) [Only if operators are present; otherwise Toll Point (TP)]

Class 5. End Office (EO) [Local central office]

Note 2: Any one center handles traffic from one center to two or more centers lower in the hierarchy. Since divestiture, these designations have become less firm.

off line. 1. In computer technology, the state or condition of a device or equipment that is not under the direct control of another device. 2. In computer technology, the status of a device that is disconnected from service.

off-line. 1. Pertaining to the operation of a functional unit when not under the direct control of the system with which it is associated. *Note 1:* Off-line units are not available for immediate use on demand by the system. *Note 2:* Off-line units may be independently operated. 2. Pertaining to equipment that is disconnected from a system, is not in operation, and usually has its main power source disconnected or turned off.

off-line recovery. The process of recovering nonprotected message traffic by use of an off-line processor or central processing unit.

off-line storage. Storage that is not under the control of a processing unit.

off-net calling. The process by which telephone calls that originate or pass through private switching systems in transmission networks are extended to stations in a public switched telephone system.

off-premises extension (OPX). An extension telephone, PBX station, or key system station located on property that is not contiguous with that on which the main telephone, PBX, or key system is located.

off-the-air. 1. In radio communications systems, pertaining to a station that is completely shut down, *i.e.,* that is not transmitting any signal, not even an unmodulated carrier. 2. In a radio station, pertaining to a particular source of modulation, such as a specific microphone, that is disconnected, *i.e.,* is no longer capable of modulating the carrier. *Note:* The carrier may continue unmodulated or it may be modulated by another signal source.

off-the-air monitoring. 1. In radio net operations, the listening, by the net-control station, to the transmissions of stations in the net, particularly to check the quality of their transmissions. *Note:* Off-the-air monitoring is usually performed during periods when the net-control station is not transmitting. 2. The listening, by a radio station, to its own transmissions by receiving the signal that has been transmitted by the transmitting antenna, to discover the quality of the signal being transmitted to other stations or being broadcast. *Note:* In off-the-air monitoring, the received signal must have traveled through the air a reasonable distance from the transmitting antenna and not be a signal that is tapped on its way to the transmitting antenna internal to the station or in the antenna transmission line, *i.e.,* the feeder. The monitoring distance should be such that direct inductive or capacitive coupling between the transmitting antenna and monitor antenna does not occur.

off-the-shelf. Pertaining to equipment already manufactured and available for delivery from stock.

OFN. *Abbreviation for* optical fiber, nonconductive. *Note:* OFN is the designation given by the National Fire Protection Association (NFPA) to interior fiber-optic cables which contain no electrically conductive component, and which are not certified for use in plenum or riser applications.

OFNP. *Abbreviation for* optical fiber, nonconductive, plenum. *Note:* OFNP is the designation given by the National Fire Protection Association (NFPA) to interior fiber-optic cables which contain no electrically conductive component, and which are certified for use in plenum applications.

OFNR. *Abbreviation for* optical fiber, nonconductive, riser. *Note:* OFNR is the designation given by the National

Fire Protection Association (NFPA) to interior fiber-optic cables which contain no electrically conductive component, and which are certified for use in riser applications.

oligarchically synchronized network. A synchronized network in which the timing of all clocks is controlled by a selected few clocks.

Omega. A global radionavigation system that enables user with special receivers to obtain position information by measuring phase difference between precisely timed signals radiated by a network of eight transmitting stations deployed worldwide. *Note:* The transmitted signals time-share transmission on frequencies of 10.2, 11.05, 11.33, and 13.6 kHz. Since the transmissions are coordinated with UTC (USNO), they also provide time reference.

omnidirectional antenna. An antenna that has a radiation pattern that is nondirectional in azimuth. *Note:* The vertical radiation pattern may be of any shape.

omnidirectional range station. A radionavigation land station in the aeronautical radionavigation service providing direct indication of the bearing (omnibearing) of that station from an aircraft.

ONA. *Abbreviation for* open network architecture.

on-board communication station. A low-powered mobile station in the maritime mobile service intended for use for internal communications on board a ship, or between a ship and its lifeboats and liferafts during lifeboat drills or operations, or for communication within a group of vessels being towed or pushed, as well as for line handling and mooring instructions.

144-line weighting: In telephony, a noise weighting used in a noise measuring set to measure line noise as it would be perceived if the line were terminated with a No. 144-receiver, or a similar instrument. *Note:* The meter scale readings are in dBrn (144-line).

144-receiver weighting. In telephony, a noise weighting used in a noise measuring set to measure noise across the receiver of an instrument equipped with a No. 144-receiver. *Note:* The meter scale readings are in dBrn (144-receiver).

one-way communication. Communication in which information is always transferred in only one preassigned direction. *Note 1:* One-way communication is not necessarily constrained to one transmission path. *Note 2:* Examples of one-way communications systems include broadcast stations, one-way intercom systems, and wireline news services.

one-way-only channel. A channel capable of transmission in only one direction, which cannot be reversed. *Synonym* unidirectional channel.

one-way operation. *Synonym* simplex operation (def. #1).

one-way reversible operation. *Synonym* half-duplex operation.

one-way trunk. A trunk between two switching centers, over which traffic may be originated from one preassigned location only. *Note 1:* The traffic may consist of two-way communications; the expression *"one way"* refers only to the origin of the demand for a connection. *Note 2:* At the originating end, the one-way trunk is known as an *"outgoing trunk"*; at the other end, it is known as an *"incoming trunk"*.

on-hook. 1. In telephony, the condition that exists when an operational telephone, or other user instrument, is not in use. *Note:* On-hook originally referred to the storage of an idle telephone receiver, *i.e.,* separate earpiece, on a hook that extended from a vertical post that supported the microphone also. The hook was mechanically connected to a switch that automatically disconnected the idle telephone from the network. 2. One of two possible signaling states, such as tone or no tone, or ground connection versus battery connection. *Note:* If on-hook pertains to one state, off-hook pertains to the other. 3. The idle state, *i.e.*, open loop, of a subscriber or PBX user loop. 4. An operating state of a communications link in which data transmission is disabled and a high-impedance, *i.e.*, open circuit, is presented to the link by the end instrument(s). *Note:* During the on-hook condition, the link is responsive to ringing signals.

on-hook signal. In telephony, of a circuit, a signal indicating a disconnect, unanswered call, or an idle condition.

on line. 1. In computer technology, the state or condition of a device or equipment that is under the direct control of another device. 2. In computer technology, the status of a device that is functional and ready for service.

on-line. 1. Pertaining to the operation of a functional unit when under the direct control of the system with which it is associated. *Note 1:* On-line units are available for immediate use on demand by the system without human intervention. *Note 2:* On-line units may not be independently operated. 2. Pertaining to equipment that is connected to a system, and is in operation.

online computer system. A computer system that is a part of, or is embedded in, a larger entity, such as a communications system, and that interacts in real or near-real time with the entity and its users.

on-premises extension. An extension telephone, PBX station, or key system station located on property that is contiguous with that on which the main telephone, PBX, or key system is located.

on-premises wiring. Customer-owned metallic or optical-fiber communications transmission lines, installed within or between buildings. *Note:* On-premises wiring may consist of horizontal wiring, vertical wiring, and backbone wiring, and may extend from the external network interface to the user work station areas. It includes the total communications wiring to transport current or future data, voice, LAN, and image information.

on-the-air. 1. In radio communications systems, pertaining to a station that is transmitting a carrier, whether or not the carrier is modulated. 2. In a radio station, pertaining to a particular source of modulation, such as a specific microphone, that is connected, *i.e.,* is capable of modulating the carrier.

open circuit. 1. In communications, a circuit available for use. 2. In electrical engineering, a circuit that contains an essentially infinite impedance. *Note:* An open circuit may be intentional, as in a switch, or may constitute a fault, as in a severed cable.

open dual bus. A dual bus in which the head-of-bus functions for both buses are at different locations.

open network architecture (ONA). In the context of the FCC's Computer Inquiry III, the overall design of a communication carrier's basic network facilities and services to permit all users of the basic network to interconnect to specific basic network functions and interfaces on an unbundled, equal-access basis. *Note:* The ONA concept consists of three integral components: (a) basic serving arrangements (BSAs), (b) basic service elements (BSEs), and (c) complementary network services.

open system. A system with characteristics that comply with specified, publicly maintained, readily available standards and that therefore can be connected to other systems that comply with these same standards.

open systems architecture. 1. The layered hierarchical structure, configuration, or model of a communications or distributed data processing system that (a) enables system description, design, development, installation, operation, improvement, and maintenance to be performed at a given layer or layers in the hierarchical structure, (b) allows each layer to provide a set of accessible functions that can be controlled and used by the functions in the layer above it, (c) enables each layer to be implemented without affecting the implementation of other layers, and (d) allows the alteration of system performance by the modification of one or more layers without altering the existing equipment, procedures, and protocols at the remaining layers. *Note 1:* Examples of independent alterations include (a) converting from wire to optical fibers at a physical layer without affecting the data-link layer or the network layer except to provide more traffic capacity, and (b) altering the operational protocols at the network level without altering the physical layer. *Note 2:* Open systems architecture may be implemented using the Open Systems Interconnection—Reference Model (OSI—RM) as a guide while designing the system to meet performance requirements. 2. Nonproprietary systems architecture.

Open Systems Interconnection (OSI). Pertaining to the logical structure for communications networks standardized by the International Organization for Standardization (ISO). *Note:* Adherence to the standard enables any OSI-compliant system to communicate with any other OSI-compliant system for a meaningful exchange of information.

Open Systems Interconnection (OSI)– Architecture. Communications system architecture that adheres to the set of ISO standards relating to open systems architecture.

Open Systems Interconnection (OSI)–Protocol Specification. The lowest level of abstraction within the OSI standards scheme. *Note:* Each OSI–Protocol Specification operates at a single layer. Each defines the primitive operations and permissible responses required to exchange information between peer processes in communicating systems to carry out all or a subset of the services defined within the OSI–Service Definitions for that layer.

Open Systems Interconnection–Reference Model (OSI–RM). An abstract description of the digital communications between application processes running in distinct systems. The model employs a hierarchical structure of seven layers. Each layer performs value-added service at the request of the adjacent higher layer and, in turn, requests more basic services from the adjacent lower layer:

Physical Layer. Layer 1, the lowest of seven hierarchical layers. The Physical layer performs services requested by the Data Link Layer. The major functions and services performed by the physical layer are: (a) establishment and termination of a connection to a communications medium; (b) participation in the process whereby the communication resources are effectively shared among multiple users, *e.g.*, contention resolution and flow control; and, (c) conversion between the representation of digital data in user equipment and the corresponding signals transmitted over a communications channel.

Data Link Layer. Layer 2. This layer responds to service requests from the Network Layer and issues service requests to the Physical Layer. The Data Link Layer provides the functional and procedural means to transfer data between network entities and to detect and possibly correct errors that may occur in the Physical Layer. *Note:* Examples of data link protocols are HDLC and ADCCP for point-to-point or packet-switched networks and LLC for local area networks.

Network Layer. Layer 3. This layer responds to service requests from the Transport Layer and issues service requests to the Data Link Layer. The Network Layer provides the functional and procedural means of transferring variable length data sequences from a source to a destination via one or more networks while maintaining the quality of service requested by the Transport Layer. The Network Layer performs network routing, flow control, segmentation/desegmentation, and error control functions.

Transport Layer. Layer 4. This layer responds to service requests from the Session Layer and issues service requests to the Network Layer. The purpose of the Transport Layer is to provide transparent transfer of data between end users, thus relieving the upper layers from any concern with providing reliable and cost-effective data transfer.

Session Layer. Layer 5. This layer responds to service requests from the Presentation Layer and issues service requests to the Transport Layer. The Session Layer provides the mechanism for managing the dialogue between end-user application processes. It provides for either duplex or half-duplex operation and establishes checkpointing, adjournment, termination, and restart procedures.

Presentation Layer. Layer 6. This layer responds to service requests from the Application Layer and issues service requests to the Session Layer. The

Presentation Layer relieves the Application Layer of concern regarding syntactical differences in data representation within the end-user systems. *Note:* An example of a presentation service would be the conversion of an EBCDIC-coded text file to an ASCII-coded file.

Application Layer. Layer 7, the highest layer. This layer interfaces directly to and performs common application services for the application processes; it also issues requests to the Presentation Layer. The common application services provide semantic conversion between associated application processes. *Note:* Examples of common application services of general interest include the virtual file, virtual terminal, and job transfer and manipulation protocols.

Open Systems Interconnection (OSI)—Service Definitions. The next lower level of abstraction below that of the OSI—Reference Model. The OSI—Service Definitions for each layer define the layer's abstract interface and the facilities provided to the user of the service independent of the mechanism used to accomplish the service.

Open Systems Interconnection (OSI)—Systems Management. In the Application Layer of the OSI—Reference Model (OSI—RM), the set of functions related to the management and status of various resources identified in all layers of the OSI—RM.

open waveguide. An all-dielectric waveguide in which electromagnetic waves are guided by a refractive index gradient so that the waves are confined to the guide by refraction or reflection from the outer surface of the guide or from surfaces within the guide. *Note 1:* In an open waveguide, the electromagnetic waves propagate, without radiation, within the waveguide, although evanescent waves coupled to internal waves may travel in the space immediately outside the waveguide. *Note 2:* Examples of open waveguides are (a) optical fibers and (b) planar waveguides in integrated optical circuits.

open wire. Conductors that are separately supported with insulators on poles or towers above the surface of the Earth. *Note 1:* Open wire conductors may be insulated or uninsulated. *Note 2:* Open wire may be used in both communication applications and power applications.

operand. An entity on which an operation is performed.

operating system. An integrated collection of routines that service the sequencing and processing of programs by a computer. *Note:* An operating system may provide many services, such as resource allocation, scheduling, input/output control, and data management. Although operating systems are predominantly software, partial or complete hardware implementations may be made in the form of firmware.

operating time. 1. The time interval between the instant of occurrence of a specified input condition to a system and the instant of completion of a specified operation. 2. In communications, computer, and information processing systems, the time interval between the instant a request for service is received from a user and the instant of final release of all facilities by the user or either of two users. 3. In communications systems conference calls, the time interval between the instant a request for service is received from one of a group of concurrent users and the instant all but one of the users have released all facilities.

operation. 1. The method, act, process, or effect of using a device or system. 2. A well-defined action that, when applied to any permissible combination of known entities, produces a new entity, *e.g.*, the process of addition in arithmetic–in adding 5 and 3 to obtain 8, the numbers 5 and 3 are the operands, the number 8 is the result, and the plus sign is the operator indicating that the operation performed is addition. 3. A program step, usually specified by a part of an instruction word, that is undertaken or executed by a computer. *Note:* Examples of operations include addition, multiplication, extraction, comparison, shift, transfer.

operational load. The total power requirements for communications facilities.

operational service period. 1. A period during which a telecommunications service remains in an operational state. *Note:* The operational state must be defined in accordance with specified criteria. 2. A performance measurement period, or succession of performance measurement periods, during which a telecommunications service remains in an operational service state. *Note:* An operational service period begins at the beginning of the performance measurement period in which the telecommunications service enters the operational service state, and ends at the beginning of the performance measurement period in which the telecommunications service leaves the operational service state.

operational service state. During any performance measurement period, a telecommunications service condition that existed when the calculated values of specified performance parameters were equal to or better than their associated outage thresholds.

operations security. [The] process denying to potential adversaries information about capabilities and/or intentions by identifying, controlling and protecting generally unclassified evidence of the planning and execution of sensitive activities.

operations system. In network management, a system that processes telecommunications management information and that supports and controls the performance of various telecommunications management functions. *Note:* An operations system performs surveillance and testing functions to support customer access maintenance.

OPSEC. *Acronym for* operations security.

optical amplifier. *See* fiber amplifier, optical repeater.

optical attenuator. In optical communications, a device used to reduce the power level of an optical signal. *Note 1:* Optical attenuators used in fiber optic communications systems may use a variety of principles for their functioning. Those using the gap-loss principle are sensitive to the modal distribution ahead of the attenuator, and should be used at or near the transmitting end, or they may introduce less loss than intended. Optical attenuators using absorptive or reflective techniques avoid this problem. *Note 2:* The basic types of optical attenuators are fixed, step-wise variable, and continuously variable.

optical axis. 1. Of a refractive or reflective optical element, the straight line that is coincident with the axis of symmetry of the surfaces. *Note:* The optical axis of a system is often coincident with its mechanical axis, but it need not be, *e.g.,* in the case of an off-axis parabolic reflector used to transmit signals to, or receive signals from, a geosynchronous satellite. *Contrast with* off-axis optical system. 2. In a lens element, the straight line which passes through the centers of curvature of the lens surfaces. 3. In an optical system, the line formed by the coinciding principal axes of the series of optical elements. 4. In an optical fiber, *synonym* fiber axis, which is the preferred term.

optical beamsplitter. *See* beamsplitter.

optical cable. *See* fiber optic cable.

optical cable assembly. *See* cable assembly.

optical cavity. A region bounded by two or more mirrors that are aligned to provide multiple reflections of lightwaves. *Note:* The resonator in a laser is an optical cavity. *In this sense, synonym* **resonant cavity.**

optical character reader (OCR). A device used for optical character recognition.

optical character recognition (OCR). The machine identification of printed characters through use of light-sensitive devices.

optical conductor. *Deprecated synonym for* optical fiber.

optical connector. A demountable device for attaching a cabled or uncabled optical fiber to another, or to an active device such as a transmitter. *Note 1:* A connector is distinguished by the fact that it may be disconnected and reconnected, as opposed to a splice, which permanently joins two fibers. *Note 2:* Optical connectors are sometimes erroneously referred to as *"couplers."* Such usage is incorrect and is to be avoided.

optical coupler. *See* directional coupler, star coupler (def.# 1), T-coupler.

optical density (OD). For a given wavelength, an expression of the transmittance of an optical element. *Note 1:* Optical density is expressed by log $(1/T)$ where T is transmittance. *Note 2:* The higher^{10} the optical density, the lower the transmittance. *Note 3:* Optical density times 10 is equal to transmission loss expressed in decibels, *e.g.*, an optical density of 0.3 corresponds to a transmission loss of 3 dB.

optical detector. A transducer that generates an output signal when irradiated with optical energy.

optical disk. A flat, circular, plastic disk coated with material on which bits may be stored in the form of highly reflective areas and significantly less reflective areas, from which the stored data may be read when illuminated with a narrow-beam source, such as a laser diode. *Note:* The bits are stored sequentially on a continuous spiral track.

optical dispersion. *See* dispersion.

optical fiber. A filament of transparent dielectric material, usually glass or plastic, and usually circular in cross section, that guides light. *Note 1:* An optical fiber usually has a cylindrical core surrounded by, and in intimate contact with, a cladding of similar geometry. *Note 2:* The refractive index of the core must be slightly higher than that of the cladding for the light to be guided by the fiber. *Synonym* lightguide.

optical fiber cable. *See* fiber optic cable.

optical fiber coating. *See* primary coating.

optical fiber, conductive. *See* OFC.

optical fiber, conductive, plenum. *See* OFCP.

optical fiber, conductive, riser. *See* OFCR.

optical fiber jacket. *See* sheath.

optical fiber link. *See* fiber optic link.

optical fiber, nonconductive. *See* OFN.

optical fiber, nonconductive, plenum. *See* OFNP.

optical fiber, nonconductive, riser. *See* OFNR.

optical fiber nuclear hardening. Design allowances made to prevent or ameliorate the effects of gamma or high-energy neutron radiation or bombardment, that causes some optical fibers to darken, increase attenuation, or depart from normal operating parameters. *Note:* Light sources, such as LEDs and lasers, and photodetectors, also need to be hardened to prevent similar malfunctions.

optical fiber transfer function. *See* transfer function.

optical fiber waveguide. *See* optical fiber.

optical filter. In the optical regime, an element that selectively transmits or blocks a range of wavelengths.

optical heterodyning. *See* optical mixing.

optical interface. In a fiber optic communications link, a point at which an optical signal is passed from one equipment or medium to another without conversion to an electrical signal.

optical isolator. A device that uses a short optical transmission path to accomplish electrical isolation between elements of a circuit. *Note 1:* The optical path may be air or a dielectric waveguide. *Note 2:* The transmitting and receiving elements of an optical isolator may be contained within a single compact module, for mounting, *e.g.,* on a circuit board. *Synonym* **optoisolator.**

optical junction. Any physical interface in a fiber optic system. *Note:* Source to fiber, fiber to fiber, fiber to detector, beam to prism (or lens), fiber to lens, lens to fiber, are examples of optical junctions.

optical line code. Sequences of optical pulses suitably structured to permit information transfer over an optical link.

optical link. An optical transmission channel, including any repeaters or regenerative repeaters, designed to connect two electronic or optoelectronic communications terminals. *Note:* An optical link is sometimes held to include the terminal optical transmitters and receivers, especially in the case of a communications link utilizing separate electronic terminals originally designed for metallic transmission, and retrofitted for optical transmission.

optically active material. A material that rotates the plane of polarization of light that passes through it.

optical mixing. Optical beating, *i.e.,* the mixing, *i.e.*, heterodyning, of two lightwaves (incoming signal and local oscillator) in a nonlinear device to produce a beat frequency low enough to be further processed by conventional electronic circuitry. *Note:* Optical mixing is the optical analog of heterodyne reception of radio signals. *Synonym* optical heterodyning.

optical multiplexing. *See* wavelength-division multiplexing.

optical path length. 1. In a medium of constant refractive index, *n*, the product of the geometric distance and the refractive index. 2. In a medium of varying refractive index, the integral of *näs*, where äs is an element of length along the path, and *n* is the local refractive index. *Note:* Optical path length is proportional to the phase shift that a lightwave undergoes along a path.

optical power. *See* radiant power.

optical power budget. In a fiber-optic communication link, the allocation of available optical power (launched into a given fiber by a given source) among various loss-producing mechanisms such as launch coupling loss, fiber attenuation, splice losses, and connector losses, in order to ensure that adequate signal strength (optical power) is available at the receiver. *Note 1:* The optical power budget is usually specified or expressed in dB. *Note 2:* The amount of optical power launched into a given fiber by a given transmitter depends on the nature of its active optical source (LED or laser diode) and the type of fiber, including such parameters as core diameter and numerical aperture. Manufacturers sometimes specify an optical power budget only for a fiber that is optimum for their equipment—or specify only that their equipment will operate over a given distance, without mentioning the fiber characteristics. The user must first ascertain, from the manufacturer or by testing, (a) the transmission losses for the type of fiber to be used, (b) the required signal strength for a given level of performance. *Note 3:* In addition to transmission loss, including those of any splices and connectors, allowance should be made for at least several dB of optical power margin losses, to compensate for component aging and to allow for future splices in the event of a severed cable. *Contrast with* optical power margin, bandwidth-limited operation.

optical power margin. In an optical communications link, the difference between (a) the optical power that is launched by a given transmitter into the fiber, less transmission losses from all causes, and (b) the minimum optical power that is required by the receiver for a specified level of performance. *Note 1:* The optical power margin is usually expressed in dB. At least several dB of optical power margin should be included in the optical power budget. *Note 2:* The amount of optical power launched into a given fiber by a given transmitter depends on the nature of its active optical source (LED or laser diode) and the type of fiber, including such parameters as core diameter and numerical aperture. *Contrast with* optical power budget.

optical receiver. A device that detects an optical signal, converts it to an electrical signal, and processes the electrical signal as required for further use.

optical regenerator. *See* optical repeater.

optical repeater. In an optical communication system, an optoelectronic device or module that receives an optical signal, amplifies it (or, in the case of a digital signal, reshapes, retimes, or otherwise reconstructs it), and retransmits it as an optical signal.

optical source. 1. In optical communications, a device that converts an electrical signal into an optical signal. *Note:* The two most commonly used optical sources are light-emitting diodes (LEDs) and laser diodes. 2. Test equipment that generates a stable optical signal for the purpose of making optical transmission loss measurements.

optical spectrum. By custom and practice, the electromagnetic spectrum between the wavelengths of the vacuum ultraviolet at 0.001 μm and the far infrared at 100 μm. *Note:* The term *"optical spectrum"* originally applied only to that region of the electromagnetic spectrum visible to the normal human eye, but is now considered to include all wavelengths between the shortest wavelengths of radio and the longest of x-rays. At this writing, no formal spectral limits are recognized nationally or internationally.

optical splitter. *See* directional coupler.

optical switch. A switch that enables signals in optical fibers or integrated optical circuits (IOCs) to be selectively switched from one circuit to another. *Note 1:* An optical switch may operate by (a) mechanical means such as physically shifting an optical fiber to drive one or more alternative fibers, or (b) electro-optic effects, magneto-optic effects, or other methods. *Note 2:* Slow optical switches, such as those using moving fibers, may be used for alternate routing of an optical transmission path, *e.g.,* routing around a fault. Fast optical switches, such as those using electro-optic or magneto-optic effects, may be used to perform logic operations.

optical system power margin. *See* power margin.

optical thickness. 1. The product of the physical thickness of an isotropic optical element and its refractive index. 2. Of an optical system, the total optical path length through all elements.

optical time domain reflectometer (OTDR). An optoelectronic instrument used to characterize an optical fiber. *Note 1:* An OTDR injects a series of optical pulses into the fiber under test. It also extracts, from the same end of the fiber, light that is scattered back and reflected back. The intensity of the return pulses is measured and integrated as a function of time, and is plotted as a function of fiber length. *Note 2:* An OTDR may be used for estimating the fiber's length and overall attenuation, including splice and mated-connector losses. It may also be used to locate faults, such as breaks.

optical transmittance. *See* transmittance.

optical transmitter. A device that accepts an electrical signal as its input, processes this signal, and uses it to modulate an opto-electronic device, such as an LED or an injection laser diode, to produce an optical signal capable of being transmitted via an optical transmission medium.

optical waveguide. Any structure having the ability to guide optical energy. *Note:* Optical waveguides may be (a) thin-film deposits used in integrated optical circuits (IOCs) or (b) optical fibers.

optimum traffic frequency. *Synonym* FOT.

optimum transmission frequency. *Synonym* FOT.

optimum working frequency. *Synonym* FOT.

optoelectronic. Pertaining to any device that functions as an electrical-to-optical or optical-to-electrical transducer, or an instrument that uses such a device in its operation. *Note 1:* Photodiodes, LEDs, injection laser diodes, and integrated optical circuit (IOC) elements are examples of optoelectronic devices commonly used in optical fiber communications. *Note 2: "Electro-optical"* is often erroneously used as a synonym.

optoisolator. *Synonym* optical isolator.

OPX. *Abbreviation for* off-premises extension.

orbit. The path, relative to a specified frame of reference, described by the center of mass of a satellite or other object in space subjected primarily to natural forces, mainly the force of gravity.

orbit determination. The process of describing the past, present, or predicted position of a satellite in terms of orbital parameters.

order of diversity. The number of independently fading propagation paths or frequencies, or both, used in diversity reception.

orderwire circuit. A voice or data circuit used by technical control and maintenance personnel for coordination and control actions relative to activation, deactivation, change, rerouting, reporting, and maintenance of communication systems and services. *Synonyms* engineering channel, engineering orderwire, service channel.

orderwire multiplex. A multiplex carrier set specifically designed for the purpose of carrying orderwire traffic, as opposed to one designed for carrying mission traffic.

ordinary ray. *See* birefringence.

organizer. *See* splice organizer.

originating user. The user that initiates a particular information transfer transaction. *Note:* The originating user may be either the source user or the destination user.

originator. *See* access originator, disengagement originator.

originator-to-recipient speed of service. *Synonym* speed of service (def. #1).

orthogonal multiplex. A method of combining two or more digital signals that have mutually independent pulses, thus avoiding intersymbol interference.

orthomode transducer. A device forming part of an antenna feed and serving to combine or separate orthogonally polarized signals.

oscillator. An electronic circuit designed to produce an ideally stable alternating voltage or current.

OSI. *Abbreviation for* Open Systems Interconnection.

OSI–RM. *Abbreviation for* Open Systems Interconnection–Reference Model.

OTAR. *Abbreviation for* over-the-air rekeying.

OTDR. *Abbreviation for* optical time domain reflectometer.

other common carrier (OCC). A communications common carrier—usually an interexchange carrier—that offers communications services in competition with AT&T and/or the established U.S. telephone local exchange carriers.

outage. A telecommunications system service condition in which a user is completely deprived of service by the system. *Note:* For a particular system or a given situation, an outage may be a service condition that is below a defined system operational threshold, *i.e.,* below a threshold of acceptable performance. *See* outage threshold.

outage duration. That period of time between the onset of an outage and the restoration of service.

outage probability. The probability that an outage will occur within a specified time period.

outage ratio. The sum of all the outage durations divided by the time period of measurement.

outage state. *See* outage.

outage threshold. For a supported performance parameter of a system, the value that establishes the minimum performance level at which the system is considered to remain in an operational state. *Note:* A measured parameter value better than the outage threshold indicates that the system is in a system operational state.

out-of-band emission. Emission on a frequency or frequencies immediately outside the necessary bandwidth which results from the modulation process, but excluding spurious emission.

out-of-band signaling. 1. Signaling that uses a portion of the channel bandwidth provided by the transmission medium, *e.g.*, the carrier channel, which portion is above the highest frequency used by, and is denied to, the speech or intelligence path by filters. *Note:* Out-of-band signaling results in a lowered high-frequency cutoff of the effective available bandwidth. 2. Signaling via a different channel (either FDM or TDM) from that used for the primary information transfer. *Contrast with* common-channel signaling, in-band signaling, out-slot signaling.

out-of-frame-alignment time. The time during which frame alignment is effectively lost. *Note:* The out-of-frame-alignment time includes the time to detect loss of frame alignment and the alignment recovery time.

outpulsing. The process of transmitting address information over a trunk from one switching center or switchboard to another.

output. 1. Information retrieved from a functional unit or from a network, usually after some processing. 2. An output state, or sequence of states. 3. Pertaining to a device, process, or channel involved in the production of data by a computer or by any of its components.

output angle. *Synonym* radiation angle.

output rating. 1. The expression of the stated power available at the output terminals of a transmitter when connected to the normal load or its equivalent. 2. Under specified ambient conditions, the expression of the power that can be delivered by a device over a long period of time without overheating.

outside plant. 1. In telephony, all cables, conduits, ducts, poles, towers, repeaters, repeater huts, and other equipment located between a demarcation point in a switching facility and a demarcation point in another switching facility or customer premises. *Note:* The demarcation point may be at a distribution frame, cable head, or microwave transmitter. 2. In DOD communications, the portion of intrabase communications equipment between the main distribution frame (MDF) and a user end instrument or the terminal connection for a user instrument.

out-slot signaling. Signaling performed in digital time slots that are not within the channel time slot. *Contrast with* out-of-band signaling.

outward dialing. *See* **direct** outward dialing.

ovality. 1. The attribute of an optical fiber, the cross section of the core or cladding of which deviates from a perfect circle. 2. In an optical fiber, the degree of deviation, from perfect circularity, of the cross section of the core or cladding. *Note 1:* The cross sections of the core and cladding are assumed to first approximation to be elliptical. Quantitatively, the ovality of either the core or cladding is expressed as $2(a-b)/(a+b)$, where a is the length of the major axis and b is the length of the minor axis. The dimensionless quantity so obtained may be multiplied by 100 to express ovality as a percentage. *Note 2:* Alternatively, ovality of the core or cladding may be expressed or specified by a tolerance field consisting of two concentric circles, within which the cross section boundaries must lie. *Synonym* noncircularity.

overfill. 1. The condition that prevails when the numerical aperture of an optical source, such as a laser, light-emitting diode, or optical fiber, exceeds that of the driven element, *e.g.*, optical fiber core. 2. The condition that prevails when the beam diameter of an optical source, such as a laser, light-emitting diode, or optical fiber, exceeds that of the driven element, *e.g.*, optical fiber core. *Note:* In optical communications testing, overfill in both numerical aperture and mean diameter (core diameter or spot size) is usually required.

overflow. 1. In telephony, the generation of potential traffic that exceeds the capacity of a communications system or subsystem. 2. In telephony, a count of telephone call attempts made on groups of busy trunks or access lines. 3. In telephony, traffic handled by overflow equipment. 4. In telephony, traffic that exceeds the capacity of the switching equipment and is therefore lost. 5. In telephony, on a particular route, excess traffic that is offered to another route, *i.e.*, an alternate route. 6. In digital computing, *synonym for* arithmetic overflow. 7. In digital communications, the condition that exists when the incoming data rate exceeds that which can be accommodated by a buffer, resulting in the loss of information.

overhead bit. Any bit other than a user information bit.

overhead communications. *See* overhead bit.

overhead information. Digital information transferred across the functional interface between a user and a telecommunications system, or between functional units within a telecommunications system, for the purpose of directing or controlling the transfer of user information or the detection and correction of errors. *Note:* Overhead information originated by the user is not considered to be system overhead information. Overhead information generated within the communications system and not delivered to the user is system overhead information. Thus, the user throughput is reduced by both overheads while system throughput is reduced only by system overhead.

overlay. 1. One of several segments of a computer program that, during execution, occupies the same area of main storage, one segment at a time. 2. To use repeatedly the same areas of internal storage during different stages of the execution of a program.

overload. A load, placed on a device or facility, that is greater than the device or facility is capable of handling, *i.e.,* capable of performing the functions for which it was designed. *Note:* Examples of overloads are (a) traffic on a communications system greater than the traffic capacity of the system, (b) for analog inputs, voltage levels above which an analog-to-digital converter cannot distinguish a change, and (c) in electrical circuits, an electrical current that will result in damage from overheating.

overload point. *Synonym* load capacity.

overmodulation. 1. The condition that prevails when the instantaneous level of the modulating signal exceeds the value necessary to produce 100% modulation of the carrier. *Note 1:* Overmodulation results in spurious emissions by the modulated carrier, and distortion of the recovered modulating signal. *Note 2:* Overmodulation in the sense of this definition is almost always considered a fault condition. 2. The condition that prevails when the mean level of the modulating signal is such that peaks in the modulating signal exceed the value necessary to produce 100% modulation of the carrier. *Note:* Overmodulation in the sense of this definition, if not excessive, is sometimes considered permissible.

override. 1. To preempt, manually or automatically, a prescribed procedure. *Note:* For example, one might manually override a prescribed course of action programmed to occur in the event of a fault. 2. In telephony, the entering of or seizure of, a busy circuit, *i.e.*, an occupied circuit, by a party other than those using the circuit. *Note:* For example, an attendant might override a circuit after a busy verification, or a user with a higher precedence level might override a circuit.

overshoot. 1. In the transition of any parameter from one value to another, the transitory value of the parameter that exceeds the final value. *Note:* Overshoot occurs when the transition is from a lower value to a higher value. When the transition is from a higher value to a lower value, and the parameter takes a transitory value that is lower than the final value, the phenomenon is called *"undershoot."* 2. The increased amplitude of a portion of a nonsinusoidal waveform, *i.e.*, signal, at the output of a nonlinear circuit, *e.g.*, a realizable amplifier, caused by the characteristics of the circuit. . *Note 1:* Overshoot causes distortion of the signal. *Note 2:* Overshoot may result from circuit design parameters that are intended to decrease the response time of the circuit. *Note 3:* The amount of overshoot in a given circuit is designed to minimize response time while maintaining distortion of the signal within acceptable limits. The absence or presence of overshoot, and if present, its magnitude, is a function of a circuit design parameter called *"damping." See illustration under* waveform. 3. The result of an unusual atmospheric, *e.g.*, ionospheric, condition that causes microwave signals to be received where they are not intended.

over-the-air rekeying (OTAR). Changing traffic encryption key or transmission security key in remote crypto-equipment by sending new key directly to the remote crypto-equipment over the communication path it secures.

over-the-horizon radar. A radar system that makes use of the atmospheric reflection and refraction phenomena to extend its range of detection beyond line of sight. Over-the-horizon radars may be either forward scatter or backscatter systems.

overtone. Of a sinusoidal wave, an integral multiple of the frequency, *i.e.,* the fundamental, of the wave, other than the fundamental itself. *Note 1:* The first overtone is twice the frequency of the fundamental, and thus corresponds to the second harmonic; the second overtone is three times the frequency of the fundamental, and thus corresponds to the third harmonic, *etc. Note 2:* Use of the term *overtone* is generally confined to acoustic waves, especially in applications related to music. *Contrast with* fundamental, harmonic.